Quality Control, Reliability, and Engineering Design

INDUSTRIAL ENGINEERING

A Series of Reference Books and Textbooks

Editor

WILBUR MEIER, JR.
Dean, College of Engineering
The Pennsylvania State University
University Park, Pennsylvania

Volume 1: Optimization Algorithms for Networks and Graphs, *Edward Minieka*

Volume 2: Operations Research Support Methodology, *edited by Albert G. Holzman*

Volume 3: MOST Work Measurement Systems, *Kjell B. Zandin*

Volume 4: Optimization of Systems Reliability, *Frank A. Tillman, Ching-Lai Hwang, and Way Kuo*

Volume 5: Managing Work-In-Process Inventory, *Kenneth Kivenko*

Volume 6: Mathematical Programming for Operations Researchers and Computer Scientists, *edited by Albert G. Holzman*

Volume 7: Practical Quality Management in the Chemical Process Industry, *Morton E. Bader*

Volume 8: Quality Assurance in Research and Development, *George W. Roberts*

Volume 9: Computer-Aided Facilities Planning, *H. Lee Hales*

Volume 10: Quality Control, Reliability, and Engineering Design, *Balbir S. Dhillon*

Additional Volumes in Preparation

Quality Control, Reliability, and Engineering Design

BALBIR S. DHILLON
Department of Mechanical Engineering
University of Ottawa
Ottawa, Ontario, Canada

MARCEL DEKKER, INC. New York and Basel

Library of Congress Cataloging in Publication Data

Dhillon, B. S.
Quality control, reliability, and engineering design.

(Industrial engineering ; v. 10)
Includes index.
1. Quality control. 2. Reliability (Engineering)
3. Engineering design. I. Title. II. Series.
TS156.D56 1984 620'.0045 85-1514
ISBN 0-8247-7278-4

MARCEL DEKKER, INC.
270 Madison Avenue, New York, New York 10016

Current printing (last digit):
10 9 8 7 6 5 4 3 2 1

PRINTED IN THE UNITED STATES OF AMERICA

This book is affectionately dedicated
to the memory of the late Bishan S. Dhillon

Preface

In recent years more attention than ever before has been paid to the quality and reliability of engineering systems during the design, manufacturing, and operation phases. Defense, aerospace, and nuclear power generation systems are the prime examples. Reasons for this increased attention include the complexity, sophistication, and size of engineering systems, use of previous untried technologies, complex mission requirements, and so on.

This leads to an increasing need for the quality control engineer, reliability engineer, and design engineer to work together closely. To achieve this goal it is essential that they have to a certain degree an understanding of each other's discipline. At present, to the author's knowledge, there is no book that covers the topics of quality control, reliability, and engineering design within the framework of a single volume. In addition, at present to gain knowledge of each other's specialties, these specialists must study various books, reports, or articles on each of the topics in question. This approach is time-consuming and rather cumbersome because of the specialized nature of the material.

This book is intended to fill the need for a single volume that considers these topics. Topics are treated in such a manner that the reader requires no prior knowledge to understand the contents. Sources for most of the material presented are given in the References. This important feature of the book will be useful to the reader who wishes to delve deeper into a certain area. In addition, the book contains over 80 examples and their solutions and exercises are offered at the end of each chapter.

The book is intended primarily for engineers, managers, undergraduate and graduate students, and others interested in the subjects of quality control, reliability, and engineering design. The book can be adopted for a variety of undergraduate, graduate, or short professional courses. For example, for a course in quality control, Chapters 4–9 and selected portions of Chapters 1–3 will be most useful. On the other hand, for a course in reliability, Chapters 9–12, 16, 4, and selected portions of Chapters 1–3 will be most beneficial. Chapters 13–18, 9, 4, and selected portions of Chapters 1–3 can be used for a course in engineering design. Similarly, Chapters 4–9, 9–12, 16, and 1–3 could be used for a course on quality control and reliability; and a reliability and engineering design course would use Chapters 9–18 and 1–4.

My experiences on many projects and in many environments, teaching and working with recognized experts in industry, academia, and research institutions in quality control, reliability, and engineering design filter through the pages of this text. I acknowledge the invisible contributions of many professionals, former and present colleagues, and students. I wish to thank my parents, brother, relatives, and friends for their interest in the project and constant encouragement throughout. Finally, I thank my wife Rosy for her painstaking typing of the entire manuscript, proofreading, preparing diagrams, and so on. I am also indebted for her constant encouragement.

Balbir S. Dhillon

Contents

Quality Control, Reliability, and Engineering Design

1
Introduction

1.1 NEED FOR QUALITY CONTROL, RELIABILITY, AND ENGINEERING DESIGN

Today engineering systems are becoming more complex to design and build because of factors such as accelerated growth in new technologies, system complexity, and size. To improve the chances that a system will perform satisfactorily during its useful life, the interest in reliability and quality control has been growing at a considerable rate in recent years. During the design phase of a system, nowadays, more attention is being paid to both these disciplines. This means that design engineers require considerable input from both quality control and reliability specialists. Furthermore, the quality control, reliability, and engineering design functions have to be integrated to achieve the end goal. To achieve such integration the design engineer and quality control and reliability specialists must have knowledge of each other's discipline. Once each of these specialists has the necessary knowledge of each other's discipline, many of their work-related difficulties will disappear or at least be reduced to a tolerable level. Furthermore, there are many advantages to having a certain degree of knowledge of work associates specialties. Some of these are better understanding of each other's work and problems, a better end product, and less time spent to comprehend each others problems.

1.2 HISTORY

This section briefly reviews the historical aspects of quality control, reliability, and engineering design. A more detailed historical review is presented in later chapters.

1.2.1 Quality Control

The origin of quality control can be traced as far back as the history of the industry itself. The condition in which mummies have been found and the quality of metals and inks produced in the remote past are examples of the high quality of some ancient products. However, not until the twelfth century were quality standards established by guilds [1].

The Industrial Revolution played an important role in the development of the quality control discipline because of the introduction of the concept of specialization of labor. And by the time of World War I full-time inspectors were being employed to inspect the output of the specialized labor. Furthermore, during the war, in 1916, C. N. Frazee of Telephone Laboratories made use of statistical methods in inspection problems. He developed the concept of the operating characteristic curve.

The development of quality control charts by Walter A. Shewhart of Western Electric, in 1924, is regarded as an important breakthrough in the history of quality control. In the period between 1924 and 1946 various people contributed to the field of quality control. In 1946, the American Society for Quality Control was formed. Since 1946 this society has played an important role in the development of the quality control field. During the period between 1946 and 1984 various journals and conference proceedings partially or fully devoted to the advancement of the field of quality control have appeared. See Chapter 4 for a further discussion.

1.2.2 Reliability

This is a new field relative to the quality control discipline. Its origin goes back to World War II. The original concept that a chain cannot be made stronger than its weakest link was introduced during the development of the V1 missile in Germany. During about the same period the need for reliability was also recognized in the United States Armed Forces. Various studies were conducted during the last five years of the 1940s. Some of the findings of these studies were as follows [2]:

1. For the maintenance of every 250 electronic tubes, approximately one technician was needed.
2. Repair and maintenance costs were approximately ten times more than the original acquisition cost in the United States Air Force.
3. A very large proportion (i.e., between two-thirds and three-fourths) of the Army equipment was either undergoing repairs or non-operational (i.e., out of commission).
4. About 70 percent of the time, the electronic equipment used by the Navy was nonoperative during maneuvers.

An ad hoc committee on reliability was formed by the United States Department of Defense in 1950. The permanent group known as the Advisory Group on the Reliability of Electronic Equipment (AGREE) resulted from the ad hoc committee and produced its report in 1957. However, the Institute of Electrical and Electronics Engineers (IEEE) and the American Society for Quality Control (ASQC) also played an important role in the development of the reliability discipline. Both these societies sponsor the journal entitled *IEEE Transactions on Reliability* and the Annual Reliability and Maintainability Symposium. Nowadays the symposium is also sponsored by seven other American professional societies such as the American Society of Mechanical Engineers, the American Society of Industrial Engineers, and the Society of Reliability Engineers. The following two journals are also devoted to the advancement of the reliability field:

1. *Microelectronics and Reliability*: an international journal, published by Pergamon Press, England.
2. *Reliability Engineering*: an international journal, Applied Science Publishers, England.

Ever since the 1950s various researchers and authors have contributed to the field of reliability engineering.

1.2.3 Engineering Design

As engineering drawings are considered to be the backbone of modern engineering design, we shall focus on the history of technical drawings. The history of engineering drawings goes back at least to ancient times, when, in 4000 B.C., the Chaldean engineer Gudea engraved upon a stone tablet the oldest surviving technical drawing: a plan view of a fortress [3].

Imhotep of Egypt, who built the first known pyramid, Saqqara, in 2650 B.C., was perhaps the first design engineer. However, the written evidence of use of technical drawings only dates back to Roman times. For example, the Roman architect Vitruvius wrote a treatise on architecture in 30 B.C.

The first book on engineering drawing in United States was published in the year 1849. The title of the book was *Geometrical Drawing* and it was published by William Minifie. Since that time a great many people have contributed to engineering design and drawings.

1.3 DEFINITIONS AND TERMS

This section gives some selected terms and definitions used in quality control, reliability, and engineering design [4–6]:

Quality control: This is a management function, whereby control of the quality of manufactured products and raw materials is exercised to prevent the manufacture of defective products.

Reliability: Reliability is the probability that an item will perform a stated function satisfactorily for a stated time period under specified conditions.

Engineering design: This is the activity in which various methods and scientific principles are used to decide the selection of materials and the placement of these materials to develop an item that fulfils specified requirements.

Hazard rate: This is defined as the rate of change of the number of units that have failed over the number of units that have survived at a certain time.

Failure: Failure is defined as the termination of a unit that carries out a specified function.

Redundancy: Redundancy is defined as the existence of more than one means for performing a stated function.

Design review: This is concerned with reviewing the design and specifications of a product from the viewpoint of performance and costs of production.

Control chart: The control chart is the chart that presents control limits.

Optimization: Optimization is the search for the solution that will generate the "maximum benefit."

Useful life: This is the length of time a unit functions with a failure rate considered to be acceptable.

Acceptance quality level (AQL): This is the maximum percentage of defective items for acceptance sampling purposes, that can be considered satisfactory as a process average.

Sample size: The sample size is defined as the number of items selected randomly from a lot to comprise a single sample.

Specification: This is defined as the detailed characteristics an item is required to meet without any difficulty.

Sample: A sample is a group of items selected randomly and usually from a lot.

Availability: This is defined as the probability that the item is functioning normally at any point in time when operated subject to specified conditions (in this case the total time considered is composed of logistic, administrative, active repair, and operating times).

1.4 SCOPE OF THE TEXT

In comparison to the engineering systems of two decades ago, present-day systems are far more complex to design and build. To improve the chances of having satisfactory system reliability and performance in the field at a minimum cost, various measures are taken at the system design stage. Furthermore, more attention is being given to reliability and quality during the design phase of a system as well as at the later stages. Therefore, it is essential for design, reliability, and quality control specialists and engineers to have knowledge of each other's specialties to produce various types of engineering systems effectively.

This book covers the topics of quality control, reliability, and engineering design within the framework of a single volume. To understand the contents, no prior knowledge of any of the three subjects presented is necessary because the book contains three introductory chapters (Chapters 4, 9, and 13) on quality control, reliability, and engineering design, and one chapter (Chapter 2) on essential mathematics.

1.5 SUMMARY

This chapter briefly introduces the subjects of quality control, reliability, and engineering design. The chapter describes the need for quality control, reliability, and engineering design as well as the need for combining these three topics into a single volume. A brief history of quality control, reliability, and engineering design is given. Selective definitions and terms used in quality control, reliability, and engineering design are presented. Finally the scope of the book is outlined.

EXERCISES

1. Discuss the importance of reliability engineering.
2. Define the following terms:
   a. Acceptance sampling
   b. Design specification
   c. Maintainability
   d. Lot-by-lot inspection
3. Describe the similarities and importance of quality control, reliability, and engineering design.
4. Define the terms "mean time to failure" (MTTF) and "mean time to repair" (MTTR).
5. Discuss the importance of engineering drawings in design.

REFERENCES

1. G. E. Hayes and H. G. Romig, *Modern Quality Control*, Collier Macmillan, London, 1977, pp. 3–9.
2. M. L. Shooman, *Probabilistic Reliability: An Engineering Approach*, McGraw-Hill, New York, 1968, pp. 12–13.
3. F. E. Giesecke, A. Mitchell, H. C. Spencer, I. L. Hill, R. D. Loving, and J. T. Dygdon, *Engineering Graphics*, Macmillan, New York, 1981.
4. J. J. Naresky, Reliability Definitions, *IEEE Trans. Reliability*, Vol. 19 (1970), pp. 198–200.
5. R. H. Lester, N. L. Enrich, and H. E. Mottley, *Quality Control for Profit*, Industrial Press, New York, 1977, pp. 317–323.
6. W. H. V. Alven, *Reliability Engineering*, Prentice-Hall, Englewood Cliffs, New Jersey, 1964, pp. 1–22.

2

Basic Mathematical Concepts

2.1 INTRODUCTION

Since mathematics has played a central role in the development of quality control, reliability, and engineering design, it is important to have an understanding of the mathematics relevant to these topics. Therefore, this chapter presents the mathematical concepts that are essential to such an understanding. Only those concepts are covered that are considered to be useful in understanding the subsequent chapters of the book. However, in some cases, the essential basic mathematical concepts are also introduced in later chapters. The main objective of this chapter is to remind engineers and others of the essential mathematical concepts rather than make them experts in mathematics.

2.2 ESSENTIAL MATHEMATICAL DEFINITIONS, LAWS, FORMULAS, AND METHODS

This section presents selected mathematical concepts. These are as follows:

2.2.1 Laws of Exponents

Some of these are as follows:

$$(mn)^k = m^k n^k \tag{2.1}$$

$$m^0 = 1 \qquad \text{for} \quad m \neq 0 \tag{2.2}$$

$$m^{k+g} = m^k m^g \tag{2.3}$$

$$m^{k-g} = \frac{m^k}{m^g} \tag{2.4}$$

where m and n are the positive numbers and k and g are the real numbers.

Example 2.1 Verify Eqs. (2.1)–(2.4) by setting $m = 3$, $n = 4$, $k = 2$, and $g = 1$.

(i) By substituting the specified values for m, n, and k in Eq. (2.1), we get

$$(3 \times 4)^2 = 3^2 4^2$$

where

$$(3 \times 4)^2 = 144$$

and

$$3^2 4^2 = (9)(16) = 144$$

(ii) Substituting $m = 3$ into Eq. (2.2) results in

$$3^0 = 1$$

(iii) From Eq. (2.3) for the given values of m, k, and g, we get

$$(3)^{2+1} = 3^2 3^1$$

where

$$(3)^{2+1} = 3^3 = 27$$

and

$$3^2 3^1 = 9 \times 3 = 27$$

(iv) Similarly, substituting the specified values for m, k, and g into Eq. (2.4) leads to

$$3^{2-1} = \frac{3^2}{3^1}$$

where

$$3^{2-1} = 3$$

and

$$3^1 = 3$$

2.2.2 Logarithmic Laws

Some of the logarithmic laws are as follows:

$$\log_b x^k = k \log_b x \tag{2.5}$$

$$\log_b \frac{x}{y} = \log_b x - \log_b y \tag{2.6}$$

$$\log_b xy = \log_b x + \log_b y \tag{2.7}$$

where b is the base—for example, the right-hand term of Eq. (2.5) reads "logarithm of x to the base b"; and x and y are positive real numbers.

Example 2.2 Assume that in Eq. (2.7), the values of x, y, and b, are 5, 8, and e (i.e., 2.718), respectively. Prove that the left-hand side of Eq. (2.7) is equal to its right-hand side.

Thus substituting the specified data into the left-hand side of Eq. (2.7), we get

$$\log_e (5)(8) = \log_e 40 = 3.6889$$

Similarly, the right-hand side of Eq. (2.7) yields

$$\log_e x + \log_e y = \log_e 5 + \log_e 8 = 1.6094 + 2.0794 = 3.6889$$

The above numerical results prove that the left-hand side and the right-hand side of Eq. (2.7) are equal.

2.2.3 Differentiation

By definition, if $g = f(y)$, then

$$\frac{dg}{dy} = \lim_{\Delta y \to 0} \frac{f(y + \Delta y) - f(y)}{\Delta y} \tag{2.8}$$

where $f(y)$ is a function of y and Δy is the finite interval.

Some of the rules of differentiation are as follows:

$$\frac{d(ky^m)}{dy} = kmy^{m-1} \tag{2.9}$$

$$\frac{d(m^v)}{dy} = m^v \ln m \frac{dv}{dy} \tag{2.10}$$

where y is a variable, v is a function of y, and k and m are constants.

Example 2.3 Differentiate with respect to y the following two functions:

$$f(y) = 5y^4 \tag{2.11}$$

$$f(y) = (1 - q)^y \tag{2.12}$$

where q is a constant.

The derivatives of the preceding two functions with respect to y are as follows:

$$\frac{df(y)}{dy} = 5(4)y^3 = 20y^3 \tag{2.13}$$

$$\frac{df(y)}{dy} = (1 - q)^y \ln(1 - q)\, 1 = (1 - q)^y(1 - y) \tag{2.14}$$

2.2.4 Definite Integration

The definite integral is defined as follows:

$$\int_a^b f(y)\, dy = [F(y)]_a^b = F(b) - F(a) \tag{2.15}$$

where $f(y)$ is an integrable function on the interval $[a,b]$ and, $F(y)$ is a function whose derivative with respect to y is equal to $f(y)$.

The resulting formulas for some of the definite integrals are as follows:

$$\int_0^t ce^{-cy}\, dy = 1 - e^{-ct} \tag{2.16}$$

$$\int_0^\infty e^{-cy}\, dy = \frac{1}{c} \tag{2.17}$$

$$\int_0^t \frac{by^{b-1}}{a} \exp\left(-\frac{y^b}{a}\right) dy = 1 - \exp\left[-\left(\frac{y^b}{a}\right)\right] \quad \text{for} \quad a,b > 0 \tag{2.18}$$

$$\int_0^\infty \exp\left[-\left(\frac{y^b}{a}\right)\right] dy = \frac{\Gamma(1/b)}{b(1/a)^{1/b}} \tag{2.19}$$

where

$$\Gamma\left(\frac{1}{b}\right) \equiv \int_0^\infty y^{(1/b-1)}e^{-y}\, dy \tag{2.20}$$

and where a, b, and c are the parameters.

Example 2.4 Evaluate

$$\int y^n \, dy \qquad \text{for} \quad n \neq -1 \tag{2.21}$$

Thus

$$\int y^n \, dy = \frac{y^{n+1}}{n+1} \tag{2.22}$$

2.2.5 Numerical Integration

It is necessary to obtain an antiderivative of the function $f(y)$ if a definite integral $\int_\alpha^\beta f(y)\,dy$ is to be evaluated by using the Fundamental Theorem of calculus. If it is not possible to obtain an antiderivative, then the numerical integration techniques are used to approximate the definite integral. Formulas of such techniques are as follows [1,2].

Simpson's formula:

$$\int_\alpha^\beta f(y)\,dy \simeq \frac{\beta-\alpha}{3k}(t_0 + 4t_1 + 2t_2 + 4t_3 + \cdots + 2t_{k-2} + 4t_{k-1} + t_n) \tag{2.23}$$

where k is the even number of equal parts of interval from $y = \alpha$ to $y = \beta$. A partition is determined by $\alpha = y_0, y_1, y_2, y_3, \ldots, y_{k-1}, y_k = \beta$. Thus, we let $t_0 = f(y_0)$, $t_1 = f(y_1), \ldots, t_k = f(y_k)$. Also, in Eq. (2.23) $(\beta - \alpha)/k$ is the size of each partition.

Trapezoidal formula:

$$\int_\alpha^\beta f(y)\,dy \simeq \frac{\beta-\alpha}{2k}(t_0 + 2t_1 + 2t_2 + 2t_3 + \cdots + 2t_{k-1} + t_k) \tag{2.24}$$

Rectangular formula:

$$\int_\alpha^\beta f(y)\,dy \simeq \frac{(\beta-\alpha)}{k}(t_0 + t_1 + t_2 + t_3 + \cdots + t_{k-1}) \tag{2.25}$$

2.2.6 Solution of Quadratic Equation

The quadratic equation is defined as follows [3]:

$$\alpha y^2 + \beta y + k = 0 \qquad \text{for} \quad \alpha \neq 0 \tag{2.26}$$

where α, β, and k are constants. Thus

$$y = \frac{-\beta \pm (\beta^2 - 4\alpha k)^{1/2}}{2\alpha} \tag{2.27}$$

The roots of the equation are

1. Equal and real for $(\beta^2 - 4\alpha k) = 0$
2. Unequal and real for $(\beta^2 - 4\alpha k) > 0$
3. Complex conjugate for $(\beta^2 - 4\alpha k) < 0$

If y_1 and y_2 are the roots of Eq. (2.26), then we can write the following two equations:

$$y_1 y_2 = \frac{k}{\alpha} \tag{2.28}$$

and

$$y_1 + y_2 = \frac{-\beta}{\alpha} \tag{2.29}$$

Example 2.5 Solve the following equation:

$$y^2 + 7y + 12 = 0 \tag{2.30}$$

In the preceding equation, the values for α, β, and k are 1, 7, and 12, respectively. Thus, substituting these values into Eq. (2.27) we get

$$\begin{aligned} y &= \frac{-7 \pm [7^2 - 4(1)(12)]^{1/2}}{2(1)} \\ &= \frac{-7 \pm (49 - 48)^{1/2}}{2} \\ &= \frac{-7 \pm 1}{2} \end{aligned}$$

Therefore,

$$y_1 = -7 + 1 = \frac{-6}{2} = -3$$

and

$$y_2 = \frac{-7 - 1}{2} = \frac{-8}{2} = -4$$

The roots of Eq. (2.30) are $y_1 = -3$ and $y_2 = -4$. In other words these two values of y satisfy Eq. (2.30).

2.2.7 Newton's Technique

This method is used to approximate the real roots of an equation [2], and it involves successive approximations. The following formula is used to approximate real roots of an equation:

$$y_{k+1} = y_k - \frac{f(y_k)}{f'(y_k)} \quad \text{for} \quad f'(y_k) \neq 0 \tag{2.31}$$

where the prime denotes differentiation with respect to y, and y_k is the value of the kth approximation. This technique is demonstrated by the example below.

Example 2.6 Approximate the real root of the following equation with the aid of Newton's method:

$$y^2 - 17 = 0 \tag{2.32}$$

As a first step, we may write

$$f(y) = y^2 - 17 \tag{2.33}$$

Differentiating function (2.33) with respect to y results in

$$\frac{df(y)}{dy} = 2y \tag{2.34}$$

By substituting Eqs. (2.33) and (2.34) into Eq. (2.31), we get

$$y_{k+1} = y_k - \frac{y_k^2 - 17}{2y_k} = \frac{y_k^2 + 17}{2y_k} \tag{2.35}$$

For $k = 1$ in Eq. (2.35) we select $y_1 = 4$ as the first approximation. Thus from Eq. (2.35) we get

$$y_2 = \frac{y_1^2 + 17}{2y_1} = \frac{(4)^2 + 17}{2(4)} = \frac{33}{8} = 4.125$$

For $k = 2$, using the above result (i.e., $y_2 = 4.125$) in Eq. (2.35) yields

$$y_3 = \frac{y_2^2 + 17}{2y_2} = \frac{(4.125)^2 + 17}{2(4.125)} = 4.1231$$

Similarly for $k = 3$, using the above result (i.e., $y_3 = 4.1231$) in Eq. (2.35) leads to

$$y_4 = \frac{y_3^2 + 17}{2y_3} = \frac{(4.1231)^2 + 17}{2(4.1231)} = \frac{34}{8.2462} = 4.1231$$

We observe from the values of y_3 and y_4 that both of them are the same. Thus the real root of Eq. (2.32) is $y = 4.1231$. To check the correctness of our end result, we substitute it back into equation (2.32):

$$y^2 - 17 = (4.1231)^2 - 17 = 0$$

This means the end result, $y = 4.1231$, is correct.

2.2.8 Combination Formula

This is defined as follows [3]:

$$\binom{m}{k} = \frac{m!}{k!(m-k)!} \tag{2.36}$$

where

$$m! = 1 \times 2 \times 3 \times 4 \times \cdots \times m \tag{2.37}$$

$$0! = 1 \tag{2.38}$$

The following notations are also used to represent the left-hand side of relationships (2.36):

1. $C(m,k)$
2. $m^C k$

Example 2.7 Compute the value of

1. 6!
2. $(5 - 3)!$

Using the formula (2.37), we get

$$6! = 1 \times 2 \times 3 \times 4 \times 5 \times 6 = 720$$

and

$$(5 - 3)! = 2! = 1 \times 2 = 2$$

Example 2.8 Compute the value of

1. $7^C 5$
2. $C(6,4)$

In the first case $m = 7$ and $k = 5$. Using both these values in the right-hand term of relationship (2.36), we get

$$\frac{m!}{k!(m-k)!} = \frac{7!}{5!(7-5)!} = \frac{1 \times 2 \times 3 \times 4 \times 5 \times 6 \times 7}{(1 \times 2 \times 3 \times 4 \times 5)(1 \times 2)} = 21$$

Similarly in the second case $m = 6$ and $k = 4$. Using both these values in the right-hand term of relationship (2.36) yields

$$\frac{m!}{k!(m-k)!} = \frac{6!}{4!(6-4)!} = \frac{1 \times 2 \times 3 \times 4 \times 5 \times 6}{(1 \times 2 \times 3 \times 4)(1 \times 2)} = 15$$

2.2.9 Laplace Transforms

The Laplace transform of a function is defined as follows:

$$\mathscr{L}[f(y)] = \int_0^\infty e^{-sy} f(y)\, dy \tag{2.39}$$

where $\mathscr{L}$ is the Laplace transform operator, $f(y)$ is a function, and s is the Laplace transform variable.

The following items are frequently used in the reliability analysis:

$$e^{-cy} \tag{2.40}$$

$$\frac{df(y)}{dy} \tag{2.41}$$

Thus the Laplace transforms of both these items with the aid of Laplace transform definition (2.39) are as follows:

$$\begin{aligned} \mathscr{L}(e^{-cy}) &= \int_0^\infty e^{-sy} e^{-cy}\, dy \\ &= \int_0^\infty e^{-(s+c)y}\, dy = \frac{1}{s+c} \end{aligned} \tag{2.42}$$

and

$$\mathscr{L}\left[\frac{df(y)}{dy}\right] = sF(s) - f(0) \tag{2.43}$$

Example 2.9 Find the Laplace transform of the following function:

$$f(y) = 1 - e^{-cy} \tag{2.44}$$

By substituting the above function into relationship (2.39), we get

$$\mathscr{L}[f(y)] = \int_0^\infty e^{-sy}[1 - e^{-cy}]\,dy$$

$$= \int_0^\infty e^{-sy}\,dy - \int_0^\infty e^{-(s+c)y}\,dy$$

$$= \left(\frac{e^{-sy}}{-s}\right)_0^\infty - \left[\frac{e^{-(s+c)y}}{-(s+c)}\right]_0^\infty$$

$$= \frac{1}{s} + \frac{1}{s+c} \qquad (2.45)$$

Example 2.10 Prove that the left-hand side of relationship (2.43) is equal to the right-hand side.

From the left-hand side of relationship (2.43) and the Laplace transform definition (2.39) we get [4]

$$\mathscr{L}[f'(y)] = \int_0^\infty e^{-sy} f'(y)\,dy$$

where the prime denotes differentiation with respect to y. This yields

$$\mathscr{L}[f'(y)] = \lim_{m \to \infty} \int_0^m e^{-sy} f'(y)\,dy \qquad (2.46)$$

Integrating the right-hand side of relationship (2.46) by parts yields

$$\mathscr{L}[f'(y)] = \lim_{m \to \infty} \left[e^{-sy} f(y)\,|_0^m + s \int_0^m e^{-sy} f(y)\,dy \right]$$

$$= \lim_{m \to \infty} \left[e^{-sm} f(m) - f(0) + s \int_0^m e^{-sy} f(y)\,dy \right]$$

$$= f(0) + s \int_0^\infty e^{-sy} f(y)\,dy \qquad (2.47)$$

The second term on the right-hand side of relationship (2.47) is the definition of the Laplace transform. Thus, from relationship (2.47) we get

$$\mathscr{L}[f'(y)] = sF(s) - f(0) \qquad (2.48)$$

where

$$F(s) = \mathscr{L}[f(y)]$$

The above resulting equation proves that the left-hand side of relationship (2.43) is equal to the right-hand side.

2.2.10 Probability Density Function

The probability density function of the continuous random variable is defined as follows:

$$f(y) = \frac{dF(y)}{dy} \tag{2.49}$$

where $f(y)$ is the probability density function, y is a continuous random variable, and $F(y)$ is the cumulative distribution function.

The cumulative distribution function is defined as follows:

$$F(y) = \int_{-\infty}^{y} f(t)\, dt \tag{2.50}$$

One should note here that the total area under the probability density function curve is always equal to unity.

The mean value of the continuous random variable is given by

$$E(Y) = \int_{-\infty}^{\infty} yf(y)\, dy \tag{2.51}$$

where $E(Y)$ is the mean value (i.e., the expected value).

Example 2.11 The probability density function $f(y)$ of the Rayleigh distribution is defined as follows:

$$f(y) = \frac{2y}{\theta} e^{-(y^2/\theta)} \qquad \text{for } \theta > 0, \qquad y \geq 0 \tag{2.52}$$

Obtain an expression for the cumulative distribution function.

Substituting the probability density function (2.52) into relationship (2.50) results in

$$F(y) = \int_0^y \frac{2t}{\theta} e^{-(t^2/\theta)}\, dt$$

$$= \frac{2}{\theta} \int_0^y te^{-(t^2/\theta)} dt$$

$$= 1 - e^{-(y^2/\theta)} \tag{2.53}$$

2.3 PROBABILITY

2.3.1 Definitions

Suppose that an event E can occur in k ways out of m equally likely ways; then the occurrence probability of event E may be defined as follows:

$$P(E) = \frac{k}{m} \tag{2.54}$$

where $P(E)$ is the occurrence probability of event E.

Similarly, the probability that the event E will not occur is given by

$$P(\bar{E}) = 1 - P(E) \tag{2.55}$$

By substituting Eq. (2.54) into Eq. (2.55) we get

$$P(\bar{E}) = 1 - \frac{k}{m}$$

$$= \frac{m-k}{m} \tag{2.56}$$

The sum of probabilities $P(E)$ and $P(\bar{E})$ is equal to unity, that is,

$$P(E) + P(\bar{E}) = 1 \tag{2.57}$$

2.3.2 Rules

Some of the rules of probability are as follows:

1. For each event B, $0 \leq P(B) \leq 1$.
2. The probability of union of mutually exclusive events, say B_1, B_2, $B_3, \ldots, B_k$, is given by

$$P(B_1 \cup B_2 \cup B_3 \cup \cdots B_k) = P(B_1) + P(B_2) + P(B_3) + \cdots + P(B_k) \tag{2.58}$$

   where k is the number of events and $\cup$ denotes the union of events.
3. The joint probability of the occurrence of k statistically independent events, say $B_1, B_2, B_3, \ldots, B_k$, is given by

$$P(B_1 B_2 B_3 \cdots B_k) = P(B_1)P(B_2)P(B_3) \cdots P(B_k) \tag{2.59}$$

   In the left-hand side of the above relationship the dot denotes the intersection of events.
4. The probability of a union of two independent events, say B_1 and B_2, is given by

$$P(B_1 \cup B_2) = P(B_1) + P(B_2) - P(B_1)P(B_2) \quad (2.60)$$

2.4 EXPONENTIAL PROBABILITY DENSITY FUNCTION

In this section the exponential probability density function is derived by using the Markov modeling technique.

To develop the Markov equations, we assume that the following rules are obeyed by the transition probabilities |5|:

1. In the time interval Δt, the probability of transition from one state to another is $\lambda\Delta t$. For example, in reliability work, λ is known as the failure rate. Thus in the transition diagram shown in Fig. 2.1, the parameter λ is the failure rate associated with states 0 and 1.
2. The transition probability of two or more transitions in time interval Δt is very small.

Thus the Markov state equations associated with Fig. 2.1 are as follows:

$$P_0(t + \Delta t) = P_0(t)(1 - \lambda\Delta t) \quad (2.61)$$

where

Δt = the time interval
λ = the constant failure rate
t = time
$P_0(t + \Delta t)$ = the probability of the system being in state 0 at time $t + \Delta t$
$P_0(t)$ = the probability of the system being in state 0 at time t
$(1 - \lambda\Delta t)$ = the probability of no failures (or occurrences) in Δt

and

$$P_1(t + \Delta t) = P_1(t) + P_0(t)\lambda\Delta t \quad (2.62)$$

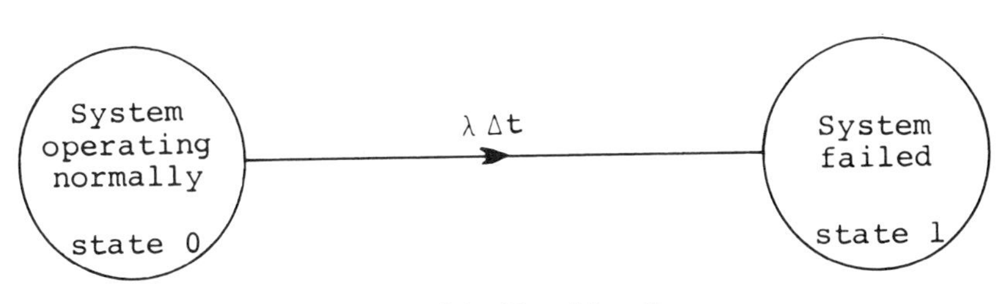

Figure 2.1 Transition diagram.

where

$P_1(t+\Delta t)=$ the probability of system being in state 1 at time $t+\Delta t$
$P_1(t)=$ the probability of system being in state 1 at time t
$\lambda\Delta t=$ the occurrence probability of transition from state 0 to state 1 in time interval Δt

Rearranging Eqs. (2.61)–(2.62) leads to

$$\frac{P_0(t+\Delta t)-P_0(t)}{\Delta t}=-\lambda P_0(t) \tag{2.63}$$

and

$$\frac{P_1(t+\Delta t)-P_1(t)}{\Delta t}=\lambda P_0(t) \tag{2.64}$$

Taking the limit of Eq. (2.6.3) as $\Delta t \to 0$ yields

$$\lim_{\Delta t\to 0}\frac{P_0(t+\Delta t)-P_0(t)}{\Delta t}=\frac{dP_0(t)}{dt}=-\lambda P_0(t) \tag{2.65}$$

and similarly

$$\lim_{\Delta t\to 0}\frac{P_1(t+\Delta t)-P_1(t)}{\Delta t}=\frac{dP_1(t)}{dt}=\lambda P_0(t) \tag{2.66}$$

At $t=0$, $P_0(0)=1$, and $P_1(0)=0$.

With the aid of relationships (2.39) and (2.43), the Laplace transforms of differential equations (2.65)–(2.66) are as follows:

$$sp_0(s)-P_0(0)=-\lambda p_0(s) \tag{2.67}$$

$$sp_1(s)-P_1(0)=\lambda p_0(s) \tag{2.68}$$

After substituting the initial conditions into Eqs. (2.67)–(2.68) we get

$$p_0(s)=\frac{1}{s+\lambda} \tag{2.69}$$

$$p_1(s)=\frac{\lambda}{s}p_0(s)=\frac{\lambda}{s}\,\frac{1}{(s+\lambda)} \tag{2.70}$$

Taking the inverse Laplace transforms of equations (2.69)–(2.70) leads to

$$P_0(t) = e^{-\lambda t} \tag{2.71}$$

$$P_1(t) = 1 - e^{-\lambda t} \tag{2.72}$$

The equation (2.72) is the same as the cumulative distribution function of the exponential distribution. To obtain the probability density function from a cumulative distribution function, we make use of the relationship (2.49). Thus using Eq. (2.72) in relationship (2.49) yields

$$f(t) = \frac{dP_1(t)}{dt} = \lambda e^{-\lambda t} \tag{2.73}$$

The preceding result represents the probability density function of the exponential distribution.

Example 12.2 The failure times of an electronic device are exponentially distributed. The constant failure rate λ of the device is 0.004 failures/hr. The device is to be operated for 400 hr. Compute the failure probability of the device.

In this example the data are specified for λ and time t as follows:

$$\lambda = 0.004 \text{ failures/hr}$$
$$t = 400 \text{ hr}$$

By substituting the preceding data into Eq. (2.72) we get

$$\begin{aligned} P_1(400) &= 1 - e^{-(0.004)(400)} \\ &= 0.7981 \end{aligned}$$

The failure probability of the device is 79.81 percent.

2.5 CRAMER'S RULE

This is concerned with solving a system of m linear equations with m unknowns [6]. A system of m linear equations with m unknowns is defined as follows:

$$\begin{array}{l} b_{11}y_1 + b_{12}y_2 + b_{13}y_3 + \cdots + b_{1m}y_m = k_1 \\ b_{21}y_1 + b_{22}y_2 + b_{23}y_3 + \cdots + b_{2m}y_m = k_2 \\ \quad\vdots \qquad\quad \vdots \qquad\quad \vdots \qquad\qquad\qquad \vdots \\ b_{m1}y_1 + b_{m2}y_2 + b_{m3}y_3 + \cdots + b_{mm}y_m = k_m \end{array} \tag{2.74}$$

where

y_i = the ith variable, $i = 1, 2, 3, 4, \ldots, m$
k_i = the ith constant, $i = 1, 2, 3, 4, \ldots, m$
b_{ij} = the (i,j) coefficient, for $i = 1$ $(i,j) = (1,1),(1,2), \ldots, (1,m)$; for $i = 2 (i,j) = (2,1), (2,2), \ldots, (2,m)$; for $i = m$ $(i,j) = (m,1), (m,2), \ldots, (m,m)$

The determinant Δ of coefficients of variables in Eq. (2.74) is

$$\Delta = \begin{vmatrix} b_{11} & b_{12} & b_{13} & \cdots & b_{1m} \\ b_{21} & b_{22} & b_{23} & \cdots & b_{2m} \\ \vdots & \vdots & \vdots & & \vdots \\ b_{m1} & b_{m2} & b_{m3} & \cdots & b_{mm} \end{vmatrix} \tag{2.75}$$

For $\Delta \neq 0$, the values of variables in Eq. (2.74) can be obtained from the following relationships:

$$y_1 = \frac{\Delta_1}{\Delta} \tag{2.76}$$

$$y_2 = \frac{\Delta_2}{\Delta} \tag{2.77}$$

$$\vdots \qquad \vdots$$

$$y_m = \frac{\Delta_m}{\Delta} \tag{2.78}$$

where Δ_i is the ith determinant of the matrix obtained by substituting the column of constants (i.e., $k_1, k_2, k_3, \ldots, k_m$) for the ith column of the coefficient matrix whose determinant is Δ; for $i = 1, 2, 3, \ldots, m$.

Example 2.13 Solve the following system of three linear equations with three unknowns by Cramer's rule:

$$\begin{aligned} 2y_1 + 4y_2 + 8y_3 &= 20 \\ 5y_1 + 10y_2 + 15y_3 &= 30 \\ 4y_1 + 5y_2 + 10y_3 &= 40 \end{aligned} \tag{2.79}$$

where y_1, y_2, and y_3 are the unknowns.

With the aid of relationship (2.75) the determinant, Δ, of coefficients of unknowns in Eqs. (2.79) is

$$\Delta = \begin{vmatrix} 2 & 4 & 8 \\ 5 & 10 & 15 \\ 4 & 5 & 10 \end{vmatrix}$$

$$= 2\begin{vmatrix} 10 & 15 \\ 5 & 10 \end{vmatrix} - 4\begin{vmatrix} 5 & 15 \\ 4 & 10 \end{vmatrix} + 8\begin{vmatrix} 5 & 10 \\ 4 & 5 \end{vmatrix}$$

$$= 2(100 - 75) - 4(50 - 60) + 8(25 - 40)$$

$$= -30$$

Similarly

$$\Delta_1 = \begin{vmatrix} 20 & 4 & 8 \\ 30 & 10 & 15 \\ 40 & 5 & 10 \end{vmatrix}$$

$$= 20\begin{vmatrix} 10 & 15 \\ 5 & 10 \end{vmatrix} - 4\begin{vmatrix} 30 & 15 \\ 40 & 10 \end{vmatrix} + 8\begin{vmatrix} 30 & 10 \\ 40 & 5 \end{vmatrix}$$

$$= 20(100 - 75) - 4(300 - 600) + 8(150 - 400)$$

$$= -300$$

$$\Delta_2 = \begin{vmatrix} 2 & 20 & 8 \\ 5 & 30 & 15 \\ 4 & 40 & 10 \end{vmatrix}$$

$$= 2\begin{vmatrix} 30 & 15 \\ 40 & 10 \end{vmatrix} - 20\begin{vmatrix} 5 & 15 \\ 4 & 10 \end{vmatrix} + 8\begin{vmatrix} 5 & 30 \\ 4 & 40 \end{vmatrix}$$

$$= 2(300 - 600) - 20(50 - 60) + 8(200 - 120)$$

$$= 240$$

$$\Delta_3 = \begin{vmatrix} 2 & 4 & 20 \\ 5 & 10 & 30 \\ 4 & 5 & 40 \end{vmatrix}$$

$$= 2\begin{vmatrix} 10 & 30 \\ 5 & 40 \end{vmatrix} - 4\begin{vmatrix} 5 & 30 \\ 4 & 40 \end{vmatrix} + 20\begin{vmatrix} 5 & 10 \\ 4 & 5 \end{vmatrix}$$

$$= 2(400 - 150) - 4(200 - 120) + 20(25 - 40)$$
$$= -120$$

Using the calculated values for Δ, Δ_1, Δ_2, and Δ_3 into Eqs. (2.76)–(2.78) we get

$$y_1 = \frac{\Delta_1}{\Delta} = \frac{-300}{-30} = 10$$

$$y_2 = \frac{\Delta_2}{\Delta} = \frac{240}{-30} = -8$$

$$y_3 = \frac{\Delta_3}{\Delta} = \frac{-120}{-30} = 4$$

The values of unknowns, y_1, y_2, and y_3 are 10, −8, and 4, respectively.

2.6 SUMMARY

This chapter briefly presents various mathematical concepts considered to be essential for understanding the subsequent chapters. Laws of exponents and logarithms are covered along with differentiation and definite integration. Three techniques of numerical integration are presented. These are the Simpson rule, the trapezoidal rule, and the rectangular formula. A formula to find roots of the quadratic equation is presented along with Newton's method for approximating the real roots of an equation. The combination formula and the definition of the Laplace transform are presented. The probability density function, the cumulative distribution function, and the expected value are defined. Some of the rules associated with probability are outlined. The exponential probability density function is derived with the aid of the Markov modeling technique. A method known as Cramer's rule is described. The method is concerned with solving a system of k linear equations with k unknowns.

EXERCISES

1. Prove by assigning the numerical values for m, n, and k that the left-hand side of Eq. (2.80) is equal to its right-hand side:

$$\left(\frac{m}{n}\right)^{1/k} = \frac{m^{1/k}}{n^{1/k}} \tag{2.80}$$

2. Determine the value of the natural logarithms for the following items:
   (i) $\log_e(e)$
   (ii) $\log_e(e^2)$
3. Differentiate the following function with respect to y:

$$f(y) = \frac{y^2 + y}{2y + 1} \qquad \text{for} \quad (2y + 1) \neq 0 \tag{2.81}$$

4. Evaluate

$$\int_0^\infty e^{-\lambda y^2}\, dy \tag{2.82}$$

   where λ is a constant.
5. Evaluate

$$\int_0^\infty y^k e^{-\lambda y}\, dy \tag{2.83}$$

   where λ and k are the constants.
6. Evaluate

$$\int_0^\infty y^k e^{-\lambda y^2}\, dy \tag{2.84}$$

7. Find the roots of the following equation:

$$4y^2 + 2y + 7 = 0 \tag{2.85}$$

8. With the aid of Newton's method approximate the real root of the following equation:

$$y^3 - 39 = 0 \tag{2.86}$$

9. Compute the value of
   1. $\binom{10}{4}$
   2. $12^{C}7$
10. Find the Laplace transform of the following function:

$$f(y) = (2 - e^{-\mu y}) \tag{2.87}$$

   where μ is a constant.
11. Find the inverse Laplace transform of the following Laplace transform:

$$\frac{1}{(s - \lambda)^k} \qquad \text{for} \quad k = 1, 2, 3, 4 \tag{2.88}$$

where s is the Laplace transform variable.

12. Find the inverse Laplace transform of the following Laplace transform:

$$\frac{1}{s^k} \quad \text{for} \quad k > 0 \tag{2.89}$$

13. Find the mean of the following probability density function:

$$f(y) = (1/a!b^{a+1})y^a e^{-y/b} \quad \text{for} \quad y > 0,\ b > 0,\ a > -1 \tag{2.90}$$

where a is the shape parameter and b is the scale parameter.

14. Solve the following system of four linear equations with four unknowns:

$$4x + 2y + 2z + 10w = 40$$
$$2x + 5y + 4z + 15w = 50$$
$$10x + 8y + 2z + 4w = 30$$
$$5x + 2y + 8z + 4w = 20$$

where x, y, z, and w are the unknowns.

15. Prove that the total area under the exponential probability density function curve is equal to unity.

REFERENCES

1. M. R. Spiegel, *Mathematical Handbook of Formulas and Tables*, McGraw-Hill, New York, 1968, pp. 95.
2. E. W. Swokowski, *Calculus with Analytic Geometry*, Prindle, Weber & Schmidt, Boston, Massachusetts, 1979, pp. 259–265.
3. W. H. V. Alven, *Reliability Engineering*, Prentice-Hall, Englewood Cliffs, New Jersey, 1964, pp. 44.
4. M. R. Spiegel, *Laplace Transforms*, McGraw-Hill, New York, 1965, pp. 15.
5. M. L. Shooman, *Probabilistic Reliability: An Engineering Approach*, McGraw-Hill, New York, 1968, pp. 229.
6. E. F. Haeussler and R. S. Paul, *Introductory Mathematical Analysis*, Reston Publishing Company, Reston, Virginia, 1980, pp. 713.

3

Economic Considerations

3.1 INTRODUCTION

In brief the main functions of engineering [1] are as follows:

1. To find solutions to problems experienced in our complex industrial-social system
2. To provide public services with highest reliability, quality, and safety at a lowest cost
3. To increase the pleasures of life

Thus any engineering design must take into consideration the economic factors that the real world is associated with. As the environments in which the engineering product has to exist become increasingly competitive and demanding, more and more attention is being given to economic aspects. Therefore, this chapter is concerned with the economics of quality control, reliability, and engineering design. The decision makers in these three areas may have to make decisions on such questions as:

1. Is it economical to improve reliability of an engineering system by introducing redundancy or improving the reliability of its individual components?
2. Which of the quality control methods are to be used at the manufacturing stage in order to ensure maximum product quality at a minimum cost?
3. What would be the production and the life cycle costs of the product under design?

The chapter is divided into four parts, on quality control, reliability, engineering design, and value analysis.

3.2 QUALITY CONTROL ECONOMICS

Like any other engineering manager, the quality control manager is also concerned with the cost of and return on services. Furthermore, the manager has to justify the cost of the quality department to the top level company management. Therefore, the manager and his or her subordinates get involved in various economics aspects of quality control program. The important aspects of quality control economics are presented in the following sections.

3.2.1 Classifications of Quality Costs

Because of conflict in objectives, the categorization of quality costs may vary from one organization to another. Therefore, there are several approaches which are practiced for recording and accounting quality costs. However, the quality costs should be classified into the following four categories [2–4]:

1. Postdelivery failure costs
2. Preventive costs
3. Costs of internal failures
4. Evaluation costs

The above four categories are described in detail as follows.

Postdelivery Failure Costs

These are those costs which have been incurred after the product was shipped to the buyer. The reasons for these costs are defective parts or products. For example, the costs associated with warranty charges, complaint adjustment, returned material, and so on are known as postdelivery failure costs.

Preventive Costs

These costs are associated with measures taken to prevent the manufacture of substandard or defective materials, components, equipment, and systems. The error prevention activities are quality planning, design review, supplier evaluation, process control, training, equipment calibration, and so on.

Costs of Internal Failures

These costs are incurred at the manufacturer's facility, in other words, prior to the shipment of the manufactured product to the customer. The reasons for these costs are that the materials, components, and systems fail to meet specified quality requirements, because of defects. More clearly, these costs are associated with repair and rework, scrap, reinspection and retest of repaired or reworked products, disposition, downtime of facilities because of defects, and so on.

Evaluation Costs

These costs are also known as appraisal costs and are associated with evaluating the condition of products from the quality requirement aspects. The evaluation costs mainly pertain to inspection-oriented activities. Nevertheless, these costs are associated with incoming material inspection, in-process inspection, assembly inspection, review and recording of data, auditing the quality system, and so on.

3.2.2 Cost Reduction Approaches

This section presents procedures [5] to reduce failure costs and to reduce appraisal costs. These two items are discussed individually as follows:

Failure Costs Reduction

This is basically a four-step approach. These steps are as follows:

1. Publicize the problems and their possible causes to make concerned groups aware.
2. Motivate concerned people to solve the existing problem; in other words create environments such that people have the desire to solve the problem in question.
3. Plan systematically and investigate the problem with the persons concerned.
4. Review the result of the action taken.

These four steps are described in detail in Ref. 3.

Appraisal Costs Reduction

According to Ref. 5, the appraisal or evaluation costs sometimes may reach 50 percent of the total costs of the quality in an organization. The techniques and approaches which will have a significant impact in improving appraisal costs are work sampling, inspection and test planning, decision analysis, evaluation (appraisal) accuracy studies, method study, and so on. All of these items are self-explanatory; nevertheless, detailed discussion of these items may be found in Ref. 5.

3.2.3 Quality Costs Utilization

There are basically two areas [5] in which the quality costs are utilized. These are as follows:

1. Strategic quality program planning
2. Programming quality improvement

Both these items are described in detail as follows.

Strategic Quality Program Planning

As each year passes, pressure to produce more reliable and safer product at a minimum cost is becoming stronger. Therefore, the strategic quality program planning is essential to fulfill these and other goals. In addition, today's competitive environments demand that the objective of the quality program must be to minimize quality costs and enhance item quality. The strategic quality planning is an ongoing process which basically is composed of the following items:

1. Appraisals
2. Making decisions
3. Taking actions

In more detail the basic steps involved in strategic planning cycle are given in Fig. 3.1.

Programming Quality Improvement

The management commitment for the improvement of quality cost and quality are discussed in the strategic quality program plan. The areas which are prime candidates for improvement can be identified with the aid of the quality cost data. For programming quality improvement it is vital:

1. To gain knowledge of the problem, recognize and systematically organize costs associated with quality
2. To perform quality performance analysis
3. To implement corrective measure; in addition, implement cost reduction programs
4. To appraise the corrective measure effort
5. To program functions in a way that will maximize utilization of personnel and monetary return
6. To budget quality activities to fulfill set goals

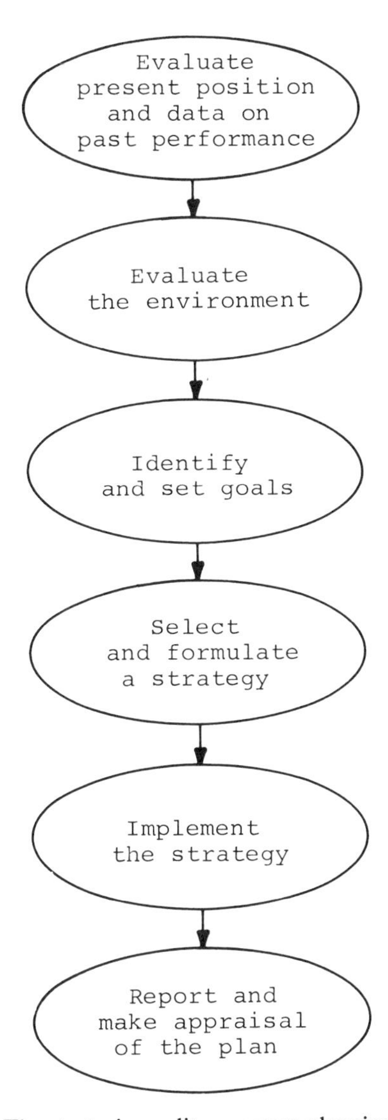

Figure 3.1 The strategic quality program planning process steps.

3.2.4 Vendor Control Quality Costs

These are of basically two types [6]:

1. Visible costs
2. Invisible costs

The visible costs are associated with acceptance sampling, warranty on items purchased from vendor, scrap and rework for which vendor is responsible, and inspection performed at the receiving end. On the other hand the hidden quality costs are incurred by the customer finding the solution to problems at the vendor facility, by the vendor itself (at its facility), and by the customer because of resulting actual or potential vendor problems.

3.2.5 Vendor Rating Program Index

This index makes use of the quality cost to appraise supplier quality cost performance. The index [6] is defined as follows:

$$QI = \frac{C_{vq} + C_p}{C_p} \tag{3.1}$$

where

QI = value of the quality cost performance index
C_{vq} = the vendor quality cost
C_p = the purchased cost

For a perfect vendor the value of the index is equal to unity since there is no vendor quality cost. For example, there would be no receiving inspection, no complaints to investigate, no defective rejections, and so on. In one example in Ref. 6 it was stated that a value of the index equal to 1.1 or more indicates that there is an immediate need for corrective measures.

Example 3.1 For the following given data calculate the value of the quality cost performance index and comment on its value:

vendor quality cost = \$2,000
purchased cost = \$50,000

Thus substituting the above given data in equation (3.1) results in

$$QI = \frac{2{,}000 + 50{,}000}{50{,}000} = 1.04$$

The value of the *QI* is equal to 1.04, which means the quality cost performance of the vendor in question is fair. According to the example given in Ref. 6 a value of *QI* between 1.000 and 1.009 was interpreted as an excellent performance; and an index value between 1.010 and 1.03 meant the performance was good.

3.2.6 Pareto Principle

In the economics of quality control the Pareto principle of maldistribution becomes very useful. Vilfredo Pareto was an Italian sociologist and economist who lived from 1848 to 1923. His principle [2] in relation to quality control work simply states that there are always a few kinds of defects in the hardware manufacture which loom large in occurrence frequency and severity. As far the economics of the quality control is concerned in either case the defects in question are costly. Therefore they are of great significance.

In other words, the same principle may be stated as that on an average about 80 percent of the costs occur due to 20 percent of defects. The advantage of the Pareto principle of maldistribution is that it helps to identify the area where one's effort is to be directed.

3.3 ECONOMICS OF RELIABILITY

As in the case of quality control, the main objective of the reliability consideration is to design, produce, and operate reliable and safe systems at the lowest life cycle cost. The system life cycle cost is simply composed of its procurement cost and its ownership cost. Some of the important aspects of reliability economics are described in the following sections.

3.3.1 Relationship Between Cost and Reliability

By spending more money on the quality and reliability aspects at the product design, development, and production stages one may increase the reliability of that product. In turn it will increase the procurement cost of that product. However, the ownership cost of the product will be lower. These relationships are depicted in Fig. 3.2.

The sum of the procurement cost and the ownership cost is called the product life cycle cost, in other words, the product acquisition cost plus the ownership cost. As can be easily visualized from Fig. 3.2, the total of the product procurement cost and the product ownership cost has a point where the life cycle cost is at its minimum. Therefore, the objective of the design engineering, reliability engineer, quality control engineer, and management should be to produce systems that have minimum life cycle cost.

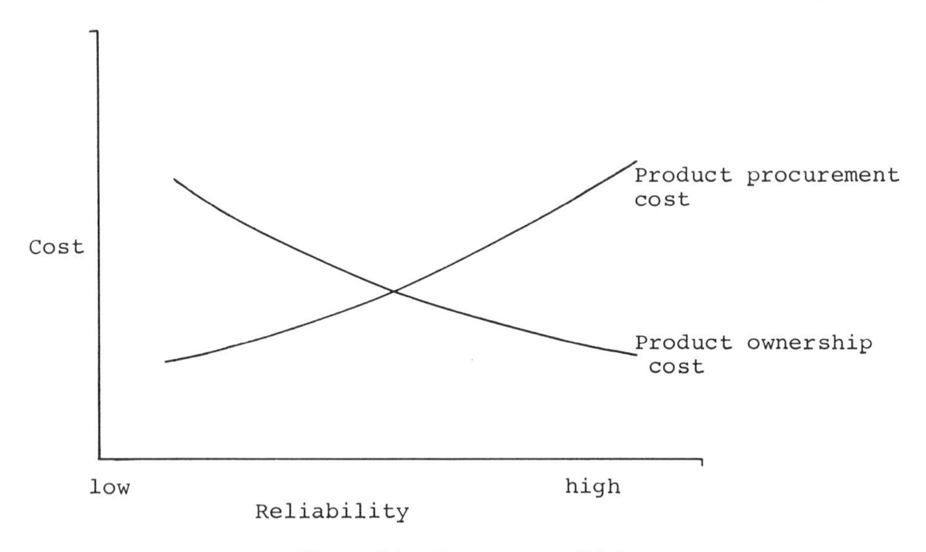

Figure 3.2 Cost versus reliability.

3.3.2 Redundancy and Cost

To improve reliability, the concept of redundancy is frequently used when designing new systems. The type and amount of redundancy varies from one situation to another. Generally speaking the redundancy is used in situations where it is highly cost effective, for example, in a system where one unit is markedly less reliable than the other units of the system as well as being cheap to purchase or produce. In any case, when introducing redundancy, careful consideration must be given to the independence of the units. Otherwise the redundant units may fail simultaneously due to a common cause, which may be costly and detrimental.

3.3.3 Life Cycle Costing

Because of present economic trend, normally consideration is given to life cycle cost when designing and developing new engineering systems, especially in the case of military systems. So far there are various approaches and mathematical models which have been developed to obtain life cycle cost of a system or a piece of equipment or a unit. According to Ref. 7 the life cycle cost of a system is the sum of research and development cost, manufacturing cost, operation and support cost; and system disposal cost. Before carrying out any type of life cost study, one should take into consideration the factors such as the availability of data constraints, uncertainties to be treated, cost analyst responsibility, objective of the estimate, accuracy and precision impact in estimates, general cost fund limitations, cost structure details requirement, and so on.

According to Ref. 9, in the development process, one should progress on the following lines:

1. Make life cycle cost (LCC) an essential component of the product development.
2. Design and develop product to LCC.
3. Control product reliability and maintainability in order to reduce costs; the major component of the product operating costs will be determined by its reliability and maintainability.
4. Make sure that the cost comparisons of alternatives are accessible to the decision makers.
5. Create environments such that at all levels of detail, the effort is being directed to minimize life cycle cost.
6. Work toward minimizing the product production cost by taking into consideration the adverse effects on its life cycle cost.

Areas of Life Cycle Cost Analysis Applications

The LCC analysis can be utilized [7] when one is evaluating many alternatives; for example, when

1. Evaluating logistic support policies
2. Evaluating alternative concepts of product maintenance
3. Determining the best production approach out of many alternatives
4. Determining the best warehouse location, distribution channel, and transportation method out of several alternatives
5. Evaluating alternative system profiles such as utilization, environmental, and operational
6. Evaluating system disposal procedures
7. Determining the best management policy out of alternative policies
8. Determining the most desirable system design configuration
9. Selecting a supplier out of many others

3.3.4 A Mathematical Model to Estimate System Development Cost

This model is taken from Ref. 10 where the system development cost C_d is given by

$$C_d = k_{ir} + k_v R^{\alpha} \tag{3.2}$$

where

$$R \equiv \frac{\ln R_{sd}}{\ln R_{id}} \tag{3.3}$$

The meanings of the symbols used in Eqs. (3.2)–(3.3) are as follows:

R_{sd} = reliability of the standard design
R_{id} = reliability of the improved design
K_{ir} = basic fixed cost which is independent of reliability
k_v = variable cost to develop a good quality and reliable system; this is composed of facilities cost, labor cost, and so on
α = a constant which takes empirical values

Equation (3.2) is subject to the following two assumptions:

1. To improve reliability due to design effort, the system development cost increases monotonically.
2. Because of the above assumption, it is assumed that as the system reliability increases, its logistic cost decreases monotonically.

3.4 ECONOMICS OF ENGINEERING DESIGN

The economic, technical, and aesthetic merits of the engineering design are vital for its commercial success. Therefore, this section is concerned with the economics of engineering design.

During the design phase of the product certain economic considerations are very important because they may be vital to the success or failure of the organization. The economic considerations concerning the market for the product under design are regarded as the most necessary economic factors.

The production and distribution of the designed product are dictated by the market requirement. Thus the designer must take into consideration the following factors when designing a new product [11]:

1. The competitive products prices
2. The percentage of the total market for the demand of the product under design
3. The size and the type of the total market
4. The price/sales relationship

and so on.

3.4.1 Design to Unit Production Cost Criteria

During the product development, design selection mainly determines the cost of production. Therefore, engineering manufacturers emphasize that the production costs must be controlled during the design phase of the product.

The following are the principal elements of an item production cost:

1. Material cost.
2. Labor cost.
3. Production overhead costs. The components of production overhead cost are the machinery depreciation cost, cost of indirect labor, services cost (fuel, electricity, etc.); cost of indirect materials such as small tools, lubricants, etc., cost of tool replacement and so on.

The design to unit production cost program is based on the following elements [9]:

1. Allocating the overall cost target among subordinate components
2. Making use of cost reduction methods during the design phase
3. Utilizing a cost-estimating approach during each design level
4. Using procedures for tracking, controlling, and reporting production costs
5. Integrating subcontractors into your own (main contractor's) effort
6. Integrating this program with the one for the life cycle cost

The above factors are described in detail in Ref. 9.

3.5 VALUE ANALYSIS AND ECONOMIC EVALUATION FORMULAS

The objective of this section is to describe briefly the value analysis and to present simple economic evaluation formulas, because both of these items are concerned directly or indirectly with quality control, reliability, and engineering design.

3.5.1 Value Analysis

This approach was originated by Lawrence D. Miles of the General Electric Company in 1947 [12]. Later the idea was picked up by other organizations. Value engineering is a systematic and creative approach used to accomplish a certain function at the lowest cost. To apply value engineering

Table 3.1 Economic Evaluation Formulas

Type	Formula[a]
Present value (single payment)	$P = AM(1+i)^{-n}$
Present value (equal payment series)	$P = Y\left[\frac{1-(1+i)^{-n}}{i}\right]$

[a] P = present value of the money
i = interest rate compounded annually
n = number of years
AM = total amount of money after n years
Y = value of each payment deposited at the end of each year

techniques, a systematic procedure known as the "job plan" is used. The job plan consists of various phases. These phases are general, information, function, creative, evaluation, investigation, and final. All of these phases are described in Ref. 13.

Economic Evaluation Formulas

The two most commonly used formulas for computing the time value of the money are given in Table 3.1.

Example 3.2 The salvage value of a piece of engineering equipment after the next ten years use is expected to be $75,000. Assume that the value of the annually compounded interest is 10 percent. Calculate the present value of that salvage money.

In this example the data are given for the following items:

$$AM = \$75{,}000, \qquad i = 10/100, \qquad n = 10 \text{ yr}$$

Thus substituting the above data in the first formula of Table 3.1 yields

$$P = 75{,}000(1+0.1)^{-10} = \$28{,}915.75$$

The present worth of the equipment salvage value is $28,915.75.

Example 3.3 A company recently purchased a computer system whose useful operating life is expected to be ten years. At the end of each of the next ten years the system will bring a revenue of $30,000. The value of annually compounded interest rate is expected to be 12 percent. Calculate the present worth of the total revenue.

In this example the data are specified for the following items:

$$Y = \$30{,}000, \qquad n = 10 \text{ yr}, \qquad i = 12/100$$

By substituting the specified data in the second formula of Table 3.1 leads to

$$P = 30{,}000 \left[\frac{1 - (1 + 0.12)^{-10}}{0.12} \right]$$

$$= \$169{,}506.69$$

Thus the present worth of the computer system total revenue will be \$169,506.69.

3.6 SUMMARY

This chapter briefly discusses the economics of quality control, reliability, and engineering design. In addition the chapter presents two commonly used economic evaluation formulas. In the quality control area, topics such as quality cost classifications, cost reduction approaches, utilization of quality costs, the Pareto principle, and vendor rating program index are covered.

The topic of reliability economics includes discussion on relationship between cost and reliability, redundancy and cost, life cycle costing, life cycle cost analysis applications, and one mathematical model to estimate system development cost.

Topics such as design to unit production cost criteria and constraints for the designer are briefly presented under the heading of economics of engineering design. The chapter contains three numerical examples along with their solutions.

The main sources of the material presented in the chapter are listed in references. These references will provide a useful service to readers if they wish to probe deeper into the subject matter.

EXERCISES

1. Describe the Pareto principle by citing examples of its application in industrial quality control work.
2. Invisible costs are one component of vendor control quality costs. Discuss these costs (invisible) in detail.
3. Describe the four types of quality costs.
4. What is meant by the term "life cycle cost of a system"?

5. Describe the term "design to cost."
6. Discuss the advantages and disadvantages of unit redundancy.
7. A machine tool procurement cost is $100,000. The expected useful life of the equipment is 15 yr. The expected value of the annual compound interest rate is 10 percent. At the end of each equipment useful life year what must be the value of revenue to break even?

REFERENCES

1. M. Kurtz, *Engineering Economics for Professional Engineers' Examinations*, McGraw-Hill, New York, 1975.
2. G. E. Hayes and H. G. Romig, *Modern Quality Control*, Bruce, Encino, California, 1977.
3. J. M. Juran and F. M. Gryna, *Quality Planning and Analysis*, McGraw-Hill, New York, 1980.
4. J. M. Juran, F. M. Gryna, and R. S. Bingham, *Quality Control Handbook*, McGraw-Hill, New York, 1979.
5. American Society for Quality Control, *Guide for Reducing Quality Costs*, Milwaukee, Wisconsin, 1977.
6. American Society for Quality Control, *Guide for Managing Vendor Quality Costs*, Milwaukee, Wisconsin, 1980.
7. B. S. Blanchard and W. J. Fabrycky, *Systems Engineering and Analysis*, Prentice-Hall, Englewood Cliffs, New Jersey, 1981.
8. J. R. Arsenault and J. A. Roberts, Eds., *Reliability and Maintainability of Electronic Systems*, Computer Science Press, Potomac, Maryland, 1980.
9. M. Robert Seldon, *Life Cycle Costing: A Better Method of Government Procurement*, Westview Press, Boulder, Colorado, 1979.
10. A. M. Hevesh, Cost of Reliability Improvement, Proceedings of the Annual Symposium on Reliability, pp. 54–61, 1969; available from the IEEE.
11. G. C. Beakley and E. G. Chilton, *Introduction to Engineering Design and Graphics*, Macmillan, New York, 1973.
12. U.S. Army Material Command, *Engineering Design Handbook, Value Engineering*, AMCP 706-104, July 1971; available from the National Technical Information Service, Springfield, Virginia.
13. A. E. Mudge, "Value Engineering: A Systematic Approach," McGraw-Hill, New York, 1971.

4

Introduction to Quality Control

4.1 INTRODUCTION

In business and industry, the importance of quality control has increased in recent times. Factors such as growing demand from customers for better quality, competition, and sophistication of product have played an important role in increasing the importance of quality control. Furthermore, our daily lives and schedules are more dependent than ever before on the satisfactory functioning of products and services—for example, automobiles, washing machines, a continuous supply of electricity, and so on. Another factor which has helped to divert more attention to the quality of products is the increasing number of lawsuits in recent years.

According to Ref. 1, roughly between 7 and 10 percent of the manufacturer's total sales revenue used up by the cost of quality control. The present quality trend facing industry may be classified [1] as follows:

1. The cost of quality has risen to a very high level.
2. Customers' quality requirements have been rising at a alarming rate.
3. Because of the above factors the present methods and practices associated with quality are rapidly becoming outmoded.

Various introductory aspects of quality control are described in the subsequent sections of this chapter.

4.2 HISTORICAL REVIEW OF QUALITY CONTROL

The origin of quality control can be traced through history as far back as the history of industry itself. For example, the inks and metals produced in the remote past were of very high quality. In addition, the condition in which the mummies have been found is another example of the excellent quality of past products.

The guilds in the twelfth century established quality standards [2]. However, it was not until the Industrial Revolution that the concept of specialization of labor came about. In other words, a worker was no longer responsible for making the whole product, but for a portion of it. This led the way to organize workers performing similar work into a group. The supervisor of the workers was made responsible for the quality of their output.

During World War I products became more complex and each supervisor had to supervise the output of a large number of workers. This introduced the problem of the supervisor controlling the quality of the output of such workers. Thus, it became necessary to examine workers' output by full-time inspectors.

The year 1916 witnessed an important development in the field of quality control when C. N. Frazee of Telephone Laboratories utilized statistical techniques for inspection problems [3]. More clearly, he made use of the operating characteristic curve. Eight years later, in 1924, Walter A. Shewhart of Bell Telephone Laboratories developed control charts to evaluate whether a specific manufacturing process was within upper and lower chance variation limits or outside such limits, i.e., out of control. This development in statistical quality control is regarded as a major milestone. According to Ref. 3, in the same decade at Bell Telephone Laboratories Harold F. Dodge and Harry G. Romig introduced the concept of acceptance sampling. In 1934 at the same establishment acceptance sampling by the technique of variables was applied. Five years later, in 1939, Romig wrote his Ph.D. thesis entitled "Allowable Average in Sampling Inspection." However, until World War II the application of statistics to quality control problems generally was not well received by the industry. The American Society for Quality Control was established in 1946.

In the 1940s and 1950s, people such as A. Wald, H.A. Freeman, A.H. Bowker, and H.P. Goode made advances to statistical quality control. A bibliography of statistical quality control was published in 1946 and updated in 1951 by Grant I. Butterbaugh.

The American Society for Quality Control (ASQC), ever since its formation, has played an important role in the development of the quality control discipline. Since 1946, the ASQC has been holding an annual conference on the topic of quality control. The following journals and conference proceedings are concerned with the subject:

1. *Quality Progress*
2. *Journal of Quality Technology*
3. *The Institute of Electrical and Electronic Engineers (IEEE) Transactions on Reliability*
4. *Annual Quality Congress Transactions of the American Society for Quality Control*

4.3 QUALITY CONTROL DEFINITIONS AND TERMS

This section presents selective definitions and terms [2,4–6] used in quality control work.

Quality: Quality may be defined as "meeting an expectation." However, there are various other definitions which are also used to describe quality. Some of them may be found in Ref. 2.

Quality control: This is a management function whereby control of the quality of manufactured item and raw materials is exercised to prevent the manufacture of defective items [4].

Quality assurance: All those systematic actions vital to provide satisfactory confidence that an item or service will fulfill defined requirements [5].

Inspection: The process of examining, testing, gauging, measuring, or otherwise comparing the item with the outlined specifications [5].

Quality system: The collective plans, events, and activities designed to ensure that an item or service will fulfill defined requirements [5].

Quality audit: Examinations, tests, and inspections conducted by quality personnel as an assistance to production in maintaining the quality of an item [6].

Control chart: A chart which possesses control limits.

Attributes sampling: Sampling in which the quality characteristic evaluated is an attribute [6].

Range: The difference between the sample's maximum and minimum values [6].

4.4 QUALITY ASSURANCE SYSTEM

This section discusses the various components of the quality assurance system which were originally outlined in Ref. 7. The objective of such a system is to maintain the necessary standard of quality. Some of the important components of the quality assurance system are as follows [7]:

1. Developing personnel
2. Monitoring quality assurance of suppliers
3. Assuring the accuracy of measuring equipment used in the quality work
4. Evaluating and controlling the quality of product in the field
5. Feeding back the quality-related information to management
6. Evaluating, planning, and controlling product quality
7. Conducting special quality studies and managing the entire quality system
8. Taking into consideration the quality and reliability needs during the product development

4.5 FACTORS WHICH AFFECT THE QUALITY OF PRODUCTS AND SERVICES

There are several basic factors which directly affect the quality of products and services. In Ref. 1, they are referred to as the nine M's. All nine of these factors are shown in Fig. 4.1.

4.6 GOALS FOR QUALITY

In many companies the attainable goals or targets for quality are established first, and then full effort is put into fulfilling such objectives. These set goals become very useful to management in planning and comparing the performance results. According to Ref. 8, in the quality function one may have either goals for control or goals for breakthrough. Goals for control are concerned with holding the quality of products or services to the present level for a specified time period—in other words maintaining the status quo. Management may have various reasons for such a decision. Some of them are as follows:

1. Insignificant complaints that demand improvement of present quality of products or services.
2. Present quality levels are quite competitive.
3. Improvements will be uneconomical.

On the other hand, the goals for breakthrough are basically concerned with improvement on the present level of quality of product or services.

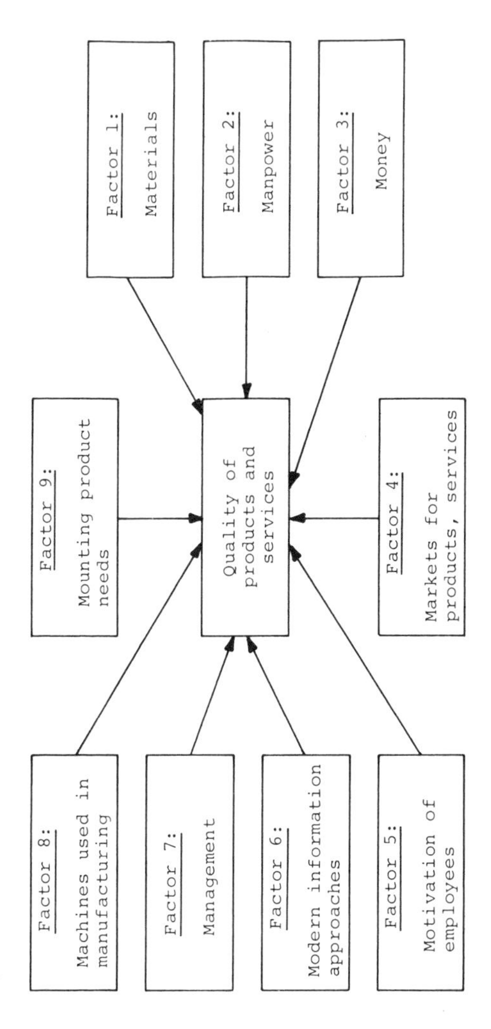

Figure 4.1 Basic factors influencing the quality of products and services.

Management has various reasons for such goals. Some of them are as follows:

1. Customers and others are dissatisfied with the present products or services.
2. Market share is being lost because of failure to compete with similar products or services provided by others.
3. The company wishes to retain or attain quality leadership.
4. The company wants to improve its image in the market.

The quality objectives or goals must be established very carefully, or the aim of such goals may not be fulfilled in a satisfactory manner. According to Juran et al. [8], the quality goals should be established by following steps such as identifying potential goals, quantifying goals, and setting priorities for goals.

4.7 SUMMARY

This chapter briefly introduces the subject of quality control. The history of quality control is summarized. Emphasis is given to the history of major developments in the statistical quality control. In addition, important quality control journals and conference proceedings are listed.

Selected quality control definitions and terms are defined. These include quality, quality assurance, quality control, quality system, control chart, inspection, and so on. Furthermore, eight major components of the quality assurance system are listed.

Factors which affect the quality of products and services are specified in Fig. 4.1. These factors are known as the 9 M's. These include manpower, money, materials, management, markets, and so on.

Finally, the chapter explains the goals for quality. Two types of goals are discussed. These are goals for control and goals for breakthrough.

EXERCISES

1. Discuss the history of statistical quality control.
2. Describe the meaning of the following terms:
   a. Quality control
   b. Control chart limits
   c. Quality audit
   d. One-hundred percent inspection
3. Are the meanings of the terms "quality assurance" and "quality control" different? If yes, describe their differences.

4. What are the general requirements for a quality program? Discuss each of them briefly.

REFERENCES

1. A. V. Feigenbaum, *Total Quality Control*, McGraw-Hill, New York, 1983, pp. 5, 24–25.
2. G. E. Hayes and H. G. Romig, *Modern Quality Control*, Collier Macmillan, London, 1977, pp. 3–9, 6–9.
3. D. H. Besterfield, *Quality Control*, Prentice-Hall, Englewood Cliffs, New Jersey, 1979, p. 2.
4. MIL-STD-109, Quality Assurance Terms and Definitions, available from the Naval Publications and Forms Center, 5801 Tabor Ave., Philadelphia, Pennsylvania, 19120.
5. ANSI/ASQC A3-1978, Quality Systems Terminology, American Society for Quality Control, 161 West Wisconsin Avenue, Milwaukee, Wisconsin 53203.
6. R. H. Lester, N. L. Enrich, H. E. Mottley, *Quality Control for Profit*, Industrial Press, New York, 1977, pp. 317–323.
7. The Quality World of Allis-Chalmers, *Quality Assurance*, Vol. 9 (1970), pp. 13–17.
8. J. M. Juran, F. M. Gryna, and R. S. Bingham, *Quality Control Handbook*, McGraw-Hill, New York, 1979, pp. 3.6–3.11.

5
Management for Quality Control

5.1 INTRODUCTION

This chapter is basically concerned with organizing for quality. The quality control function can only be performed effectively and economically if it is organized and managed properly. A properly organized quality control organization will help to simplify communication, reduce conflicts in responsibilities and activities, increase the number of satisfied customers, reduce costs, and so on. In a capsule form, a well-organized quality control organization will play an instrumental role in producing an item of acceptable quality at competitive prices. Furthermore, in today's competitive environments, any company with poorly organized quality control function will not be in business for a very long time.

According to Refs. 1 and 2, organizing for quality primarily involves the following tasks:

1. Identifying quality-related tasks
2. Assigning responsibility for accomplishing quality-related tasks to concerned company groups or departments, or to external bodies
3. Within departments breaking down the overall work into elements or jobs
4. For each job, outlining authority and responsibility
5. Defining the interrelationships among jobs
6. To accomplish the quality mission, "composing" the work of company groups or departments and external bodies

This chapter explores various areas or organizing for quality control. In addition, the topic of quality auditing is briefly discussed.

5.2 GENERAL ORGANIZATIONAL CHARTS AND METHODS

This section is concerned with general organizational charts. The main intent of introducing these charts here is to make the reader familiar with the subject. Thus, the section summarizes the basic methods of organization and associated areas.

5.2.1 Advantages of the Organizational Chart

Some of the reasons for having an organizational chart are as follows:

1. It identifies authority and fundamental relationships.
2. It assigns responsibilities.
3. It identifies strong or weak control.
4. It simplifies the management function.
5. It serves as a reference document.
6. It improves communication channels.

5.2.2 Useful Guidelines for Planning an Organization

This section lists a number of guidelines useful for planning an organization. Some of these are as follows:

1. Try to structure the organization in such a way that the organizational levels are kept at a minimum.
2. The person delegating responsibility is accountable.
3. Design the organization in such a way so that a person receives orders from only one superior authority.
4. Make sure that the lines of responsibility and authority are clear throughout the organizational structure.
5. Make sure that there is no overlap between the functional and managing duties.
6. Make sure that the delegation of authority and responsibility is close to the point of action.
7. Each component of the organization must be designed so that it accomplishes its own objective as well as that of its parent organization.

5.2.3 Organizational Procedures

This section presents selective organizational procedures used to organize activities of a company. The procedures described in this section are as follows:

1. Organization by function
2. Organization by project
3. Matrix method

The application of these methods depends on factors such as skills of company manpower, product, location of the plant, and company policy. All these methods are described below separately.

Organization by Function

This approach is widely used to organize activities of a company. This method calls for separating work according to discipline or subject—for example, organizing so that the electrical, mechanical, and hydraulic work is performed in separate groups or departments. Usually, large research organizations and organizations with long-term projects favor this approach. Some of the advantages of this method are elimination of facility duplication, more uniform products, consistent policy, even distribution of work load, etc. Similarly, some of the disadvantages are slower work flow, difficulty in shifting personnel, and so on.

Organization by Project

This is another widely used organizational approach. The project organization is the nonpermanent organizational structure established to accomplish a specific objective. In this way a company can group the appropriate talent to accomplish the specified project within defined limits. Usually after the completion of the project the associated personnel are assigned to the new project.

This type of organizational approach is favored by small and medium-size companies as well as those with short-term jobs.

Some of the advantages of the project approach are a framework for team effort, attention focused on a single project, faster work flow, better coordination of large projects, and a useful instrument for product specialization.

Similarly, some of the disadvantages of the approach are work facilities duplication, less uniform product, and inconsistent policy.

Matrix Method

Originally this method was practiced, in the 1950s and 1960s, by small and medium-size companies because they were too small to utilize the project approach. Basically the matrix approach is the result of combining both the functional and project methods. In this approach the personnel associated with a specific project report both to the management of their permanent departments and to the project manager. In fact, the project personnel are on loan to the project chief.

5.3 QUALITY CONTROL MANUAL AND ORGANIZATIONAL CHARTS

This section briefly discusses the quality control manual and organizational charts.

5.3.1 Quality Control Manual

This serves as the backbone of the quality control program. It contains various types of information concerning the quality control program of a company. There are various advantages of having such a manual. Some of them are as follows:

1. It becomes useful when making quality-related decisions.
2. It serves as a reference document.
3. It can be used as a textbook when training quality personnel.
4. It helps in continuity of operations of the quality control organization despite the personnel turnover.

Usually, the manual contains information on items such as the following:

1. Responsibilities
2. Statistical methodology
3. Personnel
4. Organizational charts
5. Quality policies and procedures
6. Vendor quality control procedure
7. Quality costs and inspection procedures
8. Measuring equipment
9. Defect prevention
10. Marketing

5.3.2 Steps for Planning the Quality Control Organizational Structure

According to Ref. 3 when planning the quality control organizational structure, the steps outlined in Fig. 5.1 are to be followed.

5.3.3 Organizing the Quality Control Program

Care must be given when organizing the quality control department and placing it in the company's organizational structure. This may depend on factors such as the philosophy of the management, size, product, and organizational structure of the company. However, factors such as those following are to be considered when organizing a quality control program in a company:

1. In organizations with less than 500 employees, it may be appropriate (with care) to combine the quality control function with another function [4]. For example, a manager is responsible for both design engineering and quality control.
2. Companies which are primarily concerned with manufacturing and have strict quality requirements should have separate quality control departments. Quality control should not be the responsibility of the production management [5]; otherwise the quality of the product may be sacrificed.
3. According to Ref. 5, the present-day trend is towards having the quality manager report directly to the upper-level management, i.e., to the vice-president, plant manager, etc. In this way the quality department will have better control over its quality-related affairs.

5.3.4 Quality Control Organizational Methods

According to Ref. 6, frequently used organizational methods for quality control are as follows:

1. Departmental
2. Advisory
3. Functional component of top management

Departmental Method

In this case there is a separate department which has the sole responsibility for quality control. Furthermore, the quality control chief reports to the general manager, who in turn reports to the president. The advantage of this

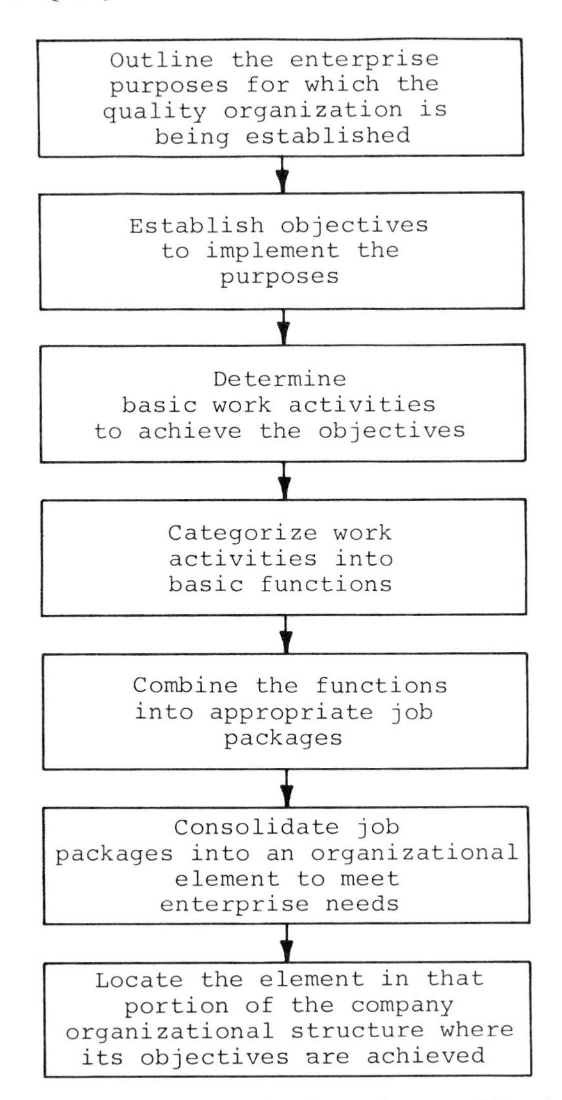

Figure 5.1 Steps to organize the quality control function.

approach is that the department chief has a definite responsibility for quality control which gives him incentive to produce effective results.

Advisory Method

This is another approach where quality control personnel such as inspection supervisors, inspectors, and others play the role of "process advisors." In other words, they do not exercise authority but act as advisors to the engineering and production departments. Furthermore, they function under the authority of a general manager, who reports directly to the president. One of the advantages of this approach is that substantial changes are not needed. Furthermore, in this way the probability of having differences between production and inspection is reduced. However, the disadvantages of this approach are the lack of authority and no definite responsibilities for the quality control advisors.

Functional Component of Top Management

This approach dictates making quality control a functional component of upper management, or, more specifically, making the position of the quality control head similar to that of the controller. In this situation, the quality control chief reports directly to the president. According to Enrick [6], the quality chief has indirect authority over the general manager and design chief with respect to quality. The significant advantage of this approach is that the quality control head can exercise his or her authority to place responsibility for quality where it belongs. In addition, the quality chief can develop the company policy for quality.

5.4 FUNCTIONS OF QUALITY CONTROL ENGINEERING

According to Charbonneau [7], the work components of the quality control engineering may be classified as shown in Fig. 5.2. These are concerned with product evaluation, new design, general quality control engineering, quality control of incoming material, process quality control, special activities, and inventory quality control. The functions outlined in Fig. 5.2. are described in detail by Charbonneau and Webster [7].

5.5 THE QUALITY CONTROL MANAGER

This section discusses the attributes, roles, and typical reasons for the failure of a quality control manager.

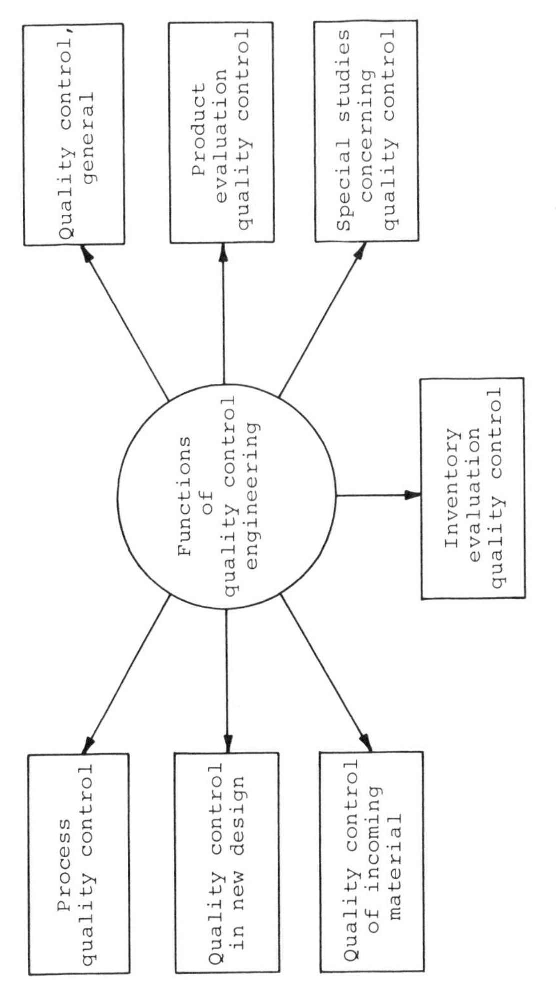

Figure 5.2 Functions of the quality control engineering.

5.5.1 Attributes of the Quality Control Manager

It is expected that the quality manager must possess certain attributes to perform his or her job effectively. Some of these are as follows:

1. Creativeness and fairness
2. Broad knowledge of engineering, production, and research
3. Competence in statistical analysis
4. Honesty and compassion
5. Performance-mindedness
6. Good health and strength of character
7. Impartiality and decision-making ability
8. Ability to listen and to motivate
9. Skill in negotiating
10. Ability to plan, organize, delegate, and sell effectively
11. Knowledge of finance and marketing functions
12. Patience and persistence
13. Communication ability
14. Enthusiasm
15. Ability to think effectively
16. Ability to utilize time effectively

5.5.2 Roles Performed by the Quality Manager

The quality manager performs various roles with regard to quality. Some of them are as follows [1]:

1. Consulting
2. Supervision of people who perform analysis
3. Supervision of inspection personnel
4. Acting as "customer's representative" with regard to quality
5. Keeping contacts with outside bodies
6. Serving as the company's "reporter and scorekeeper" with respect to the state of the quality function
7. Quality planning and coordinating

5.5.3 Reasons for the Failure of Quality Managers

There are various reasons why quality managers fail to perform their job effectively. Some of them are shown in Fig. 5.3.

5.6 THE PROCESS ENGINEER

The process engineer is accountable to the quality control manager. The responsibilities and the characteristics of the process engineer are as follows [4].

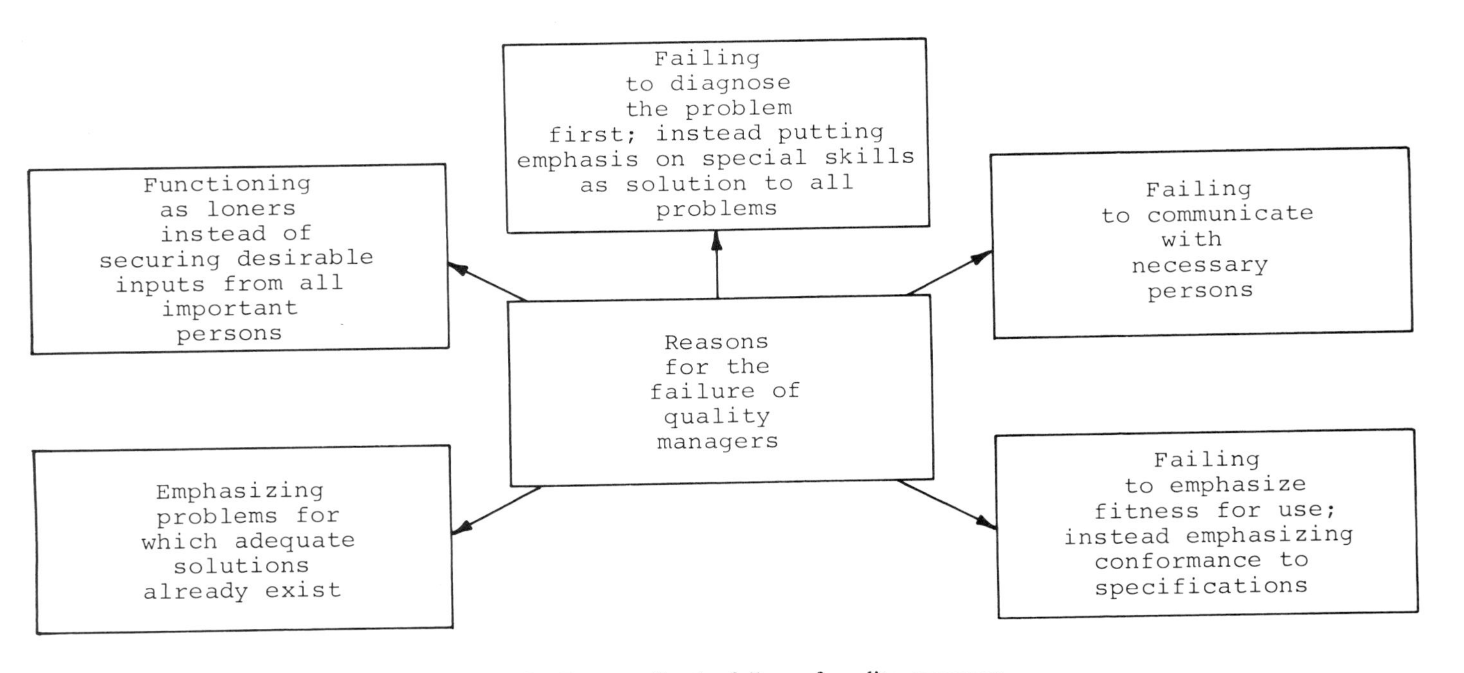

Figure 5.3 Reasons for the failure of quality managers.

5.6.1 Responsibilities of the Process Engineer

Some of the main responsibilities of the process engineer are as follows:

1. Conducts failure analysis
2. Conducts studies concerning process capability
3. Develops and maintains quality planning including tool and gauge control, audits, specifications, calibration, and so on
4. Provides training to technicians concerned with process
5. Performs safety audits of quality control
6. Makes process control recommendations
7. Provides standards for inspecting quality visually

5.6.2 Desirable Personality Traits of a Process Engineer

An effective process engineer possesses certain qualities. Some of them are as follows:

1. Creativity and persistence
2. Ability to give prompt response (with an acceptable solution) to the problem whenever it arises
3. Enthusiasm
4. Willingness to accept challenge whenever it arises
5. Capability for further growth
6. Ability to deal with people (i.e., concern for people)

5.7 QUALITY AUDITING

The quality audits are performed to find out whether [4]

1. The end product satisfies the pertinent quality specifications
2. The equipment and machinery are functioning according to expectation
3. Operators carry out their duties according to specified quality plans

For the success of the quality control system, special care must be given to the auditing of tool and gauge, testing, quality plans, specifications, and drawings.

5.7.1 Guidelines for Auditing

This section presents various guidelines which become useful in auditing. Some of these are as follows [5]:

1. Make sure that the auditing schedule is maintained.

2. Perform audits by using checklists.
3. Make sure that no one individual performs quality audits for more than half a year at a time.
4. Record the audit results.
5. Carry out audits without prior announcements.
6. Audit all the workshifts (i.e., day, night, etc.).
7. Distribute the results of auditing to all concerned people.
8. Take follow-up actions, to make sure that necessary measures are taken to correct the deficiencies found during the audits.
9. Select a person for auditing who is unbiased.
10. Do not conduct audits as if you are trying to catch someone performing wrongly.
11. Avoid making deals with people when auditing.

5.7.2 Advantages of Auditing

There are various advantages of auditing. Some of them are as follows [5]:

1. Helps to reduce the complaints received from customers
2. Causes problem areas to surface
3. Helps stabilize the levels of quality
4. Finds out if the company objectives are fulfilled
5. Determines the quality system efficiency
6. Helps to predict the reaction of the buyer toward quality
7. The documentation associated with auditing becomes useful to assure customer regarding quality
8. Determines if the customer specifications are met satisfactorily
9. Helps to improve cost effectiveness
10. Helps to show effects of changes (i.e., changes in quality planning and other changes initiated by the management)
11. Generates useful data which can be used for various purposes by the quality assurance management
12. Helps to measure the company policy implementation effectiveness
13. Measures the effectiveness of persons with respect to quality plans' implementation

5.8 SUMMARY

This chapter describes various aspects of quality control. The chapter begins by describing the tasks associated with organizing for quality. The next topics discussed in the chapter are organizational charts and methods. The reasons for having organizational charts and guidelines for planning an organization are presented. In addition, three commonly used methods in organizing are described. These are organization by project, organization by function, and the matrix approach.

The quality control manual and quality control organizational charts are the other two subjects which are discussed in the chapter. The advantages of having a quality control manual and its essential components are listed. The quality control organizational methods such as departmental, advisory, and functional component of top management are briefly discussed.

Another topic of interest, the functions of quality control engineering, is also covered in the chapter. Seven functions are listed.

Three areas directly concerning quality control managers are briefly described. These are the qualities, roles, and reasons for the failure of such managers. The next topic covered, the process engineer, is also associated with the quality personnel. Responsibilities and desirable personality traits of a process engineer are listed.

Quality auditing is the last topic of the chapter. The objectives of quality auditing are outlined. In addition, guidelines for auditing and its advantages are listed.

EXERCISES

1. What are the reasons for having an organizational chart?
2. Describe the matrix approach used to set up an organizational structure.
3. What type of information should be included in a quality control manual?
4. List the advantages of having the quality control manual.
5. What are the methods used to organize the quality control program?
6. Describe the functions of a quality control department.
7. What are the attributes of a quality control manager?
8. Describe the responsibilities of a quality control manager.
9. What are the duties of a process engineer?
10. Describe at least ten benefits of auditing.

REFERENCES

1. J. M. Juran, F. M. Gryna, and R. S. Bingham, *Quality Control Handbook*, McGraw-Hill, New York, 1974, pp. 7.1–7.2.
2. J. M. Juran and F. M. Gryna, *Quality Planning and Analysis: From Product Development Through Use*, McGraw-Hill, New York, 1980.
3. A. V. Feigenbaum, *Total Quality Control*, McGraw-Hill, New York, 1983, p. 163.

4. R. H. Lester, N. L. Enrick, and H. E. Mottley, *Quality Control for Profit*, Industrial Press, New York, 1977, pp. 5–6.
5. G. E. Hayes and H. G. Romig, *Modern Quality Control*, Collier-Macmillan, London, 1977, pp. 439–492.
6. N. L. Enrick, *Quality Control and Reliability*, Industrial Press, New York, 1972, pp. 147–150.
7. H. C. Charbonneau and G. L. Webster, *Industrial Quality Control*, Prentice-Hall, Englewood Cliffs, New Jersey, 1978, pp. 200–204.

6

Quality Costs and Procurement Quality Control

6.1 INTRODUCTION

This chapter describes two important aspects of quality control: quality costs and procurement quality control.

Quality costs play a central role in management decision making. In business terms, quality costs provide the economic common denominator through which people associated with quality and management can interact effectively. Furthermore, in monetary terms, the costs are useful for measuring the effectiveness of the quality department, quality planning, etc. In addition, the department efficiency can only be evaluated through costing and control. Knowledge of efficiency is important to management as the profit margin is a function of efficiency. Accurate profit margins and cost terms can only be determined from accurate cost data. Thus the availability of such data is essential.

Another important aspect of quality control is procurement quality control. Today it is quite unlikely that for engineering products sold in the market the manufacturer of such products has also produced all the components and raw materials used in them. Usually many components and materials are procured from many outside manufacturers and suppliers. For example, the manufacturer of television sets usually purchase items such as tubes, integrated circuits, transistors, and resistors from outside agencies or manufacturers. Thus, it is essential to give proper care to the quality control of procured items.

Quality costs and procurement quality control are discussed separately in the subsequent sections of this chapter.

6.2 QUALITY COSTS

According to Ref. 1, in several American manufacturing organizations the costs of quality at the beginning of the 1970s were 10 percent of the sales income. It was predicted that if corrective measures are not taken, the quality costs might reach 15 percent of the sales dollar in the early years of the 1980s. Thus, this section describes various important aspects of the quality costs.

6.2.1 Types of Quality Costs

Quality costs [2,3] can be divided into five categories as shown in Fig. 6.1. Components of each classification are described below.

Classification I

This classification includes all those costs associated with internal failures, in other words, the costs associated with materials, components, and products and other items which do not satisfy quality requirements.

Furthermore, these are those costs which occur before the delivery of the product to the buyer. These costs are associated with things such as the following:

1. Scrap
2. Failure analysis studies
3. Testing
4. In-house components and materials failures
5. Corrective measures

Classification II

This classification is concerned with prevention costs. These costs are associated with actions taken to prevent defective components, materials, and products. Prevention costs are associated with items such as the following:

1. Evaluating suppliers
2. Calibrating and certifying inspection and test devices and instruments
3. Receiving inspection
4. Reviewing designs
5. Training personnel
6. Collecting quality-related data
7. Coordinating plans and programs

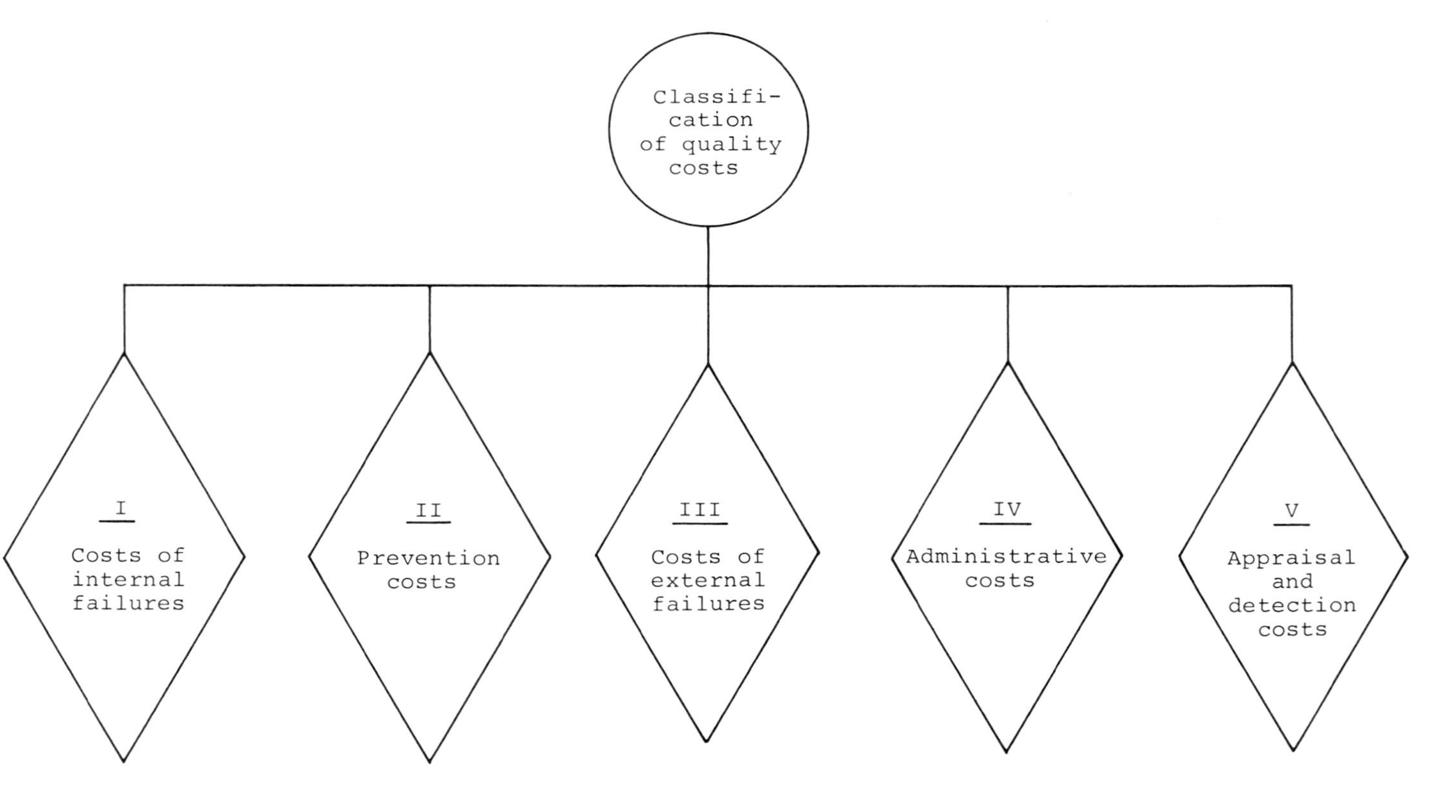

Figure 6.1 Classifications of quality costs.

8. Implementing and maintaining sampling plans
9. Preparing reliability demonstration plans

Classification III

Under this classification are costs associated with external failures—in other words, costs due to defective products shipped to the buyers. These costs are associated with item such as the following:

1. Investigation of customer complaints
2. Liability
3. Repair
4. Failure analysis
5. Warranty charges
6. Replacement of defective items

Classification IV

This category includes all the administrative-oriented costs—for example, costs associated with the following:

1. Reviewing contracts
2. Preparing proposals
3. Performing data analysis
4. Preparing budgets
5. Forecasting
6. Management
7. Clerical

Classifications V

This category includes costs associated with detection and appraisal. The principal components of such costs are as follows:

1. Cost of testing
2. Cost of inspection (i.e., in-process, source, receiving, shipping, and so on)
3. Cost of auditing

6.2.2 An Approach to Evaluating Quality Costs

This approach is based on the following index [4]:

$$\theta = \frac{a}{b}(100) + (100) \qquad (6.1)$$

where θ is known as the quality cost index, a denotes the quality costs, and b denotes the value of output.

When the value of $\theta = 100$, it means that there is no defective output. Thus no money is spent to make quality checks. However, in a real-life situation, a value of $\theta = 105$ can readily be achieved. In plants where the quality costs are ignored, the value of this index commonly varies between 110 and 130.

A practical six-step approach [4] to estimating the value of θ is shown in Figure 6.2.

6.2.3 Ways in Which the Quality Organization Increases the Company Profit

From Ref. 3 some of the ways the quality organization increases the company profit are as follows:

1. By performing the analysis of the complaints received from customers and then taking the necessary measures
2. By reducing the time spent for repair
3. Through process control, reducing the severity of defects
4. By using sampling plans and control charts
5. By assuring the achievement of planned quality
6. By performing failure analyses

6.2.4 Benefits of Quality Costs

This section presents the benefits of the quality costs. Some of them are as follows:

1. Serves as a useful tool in budgeting
2. Becomes useful in predicting the performance of the plant
3. Serves as a useful tool in determining the action priorities
4. Serves as a useful tool when pricing products
5. Serves as a useful tool to determine problem areas
6. Serves as a useful tool when bidding on project

6.3 PROCUREMENT QUALITY CONTROL

As stated earlier, the manufacturer of a product usually purchases many components and materials from outside suppliers. According to Hayes and Romig [3], on an average over 50 percent of revenue from the sales of products in American companies is accounted for by components, materials, supplies, and so on, purchased from outside companies. Thus to produce a product with acceptable quality level, the quality of purchased components

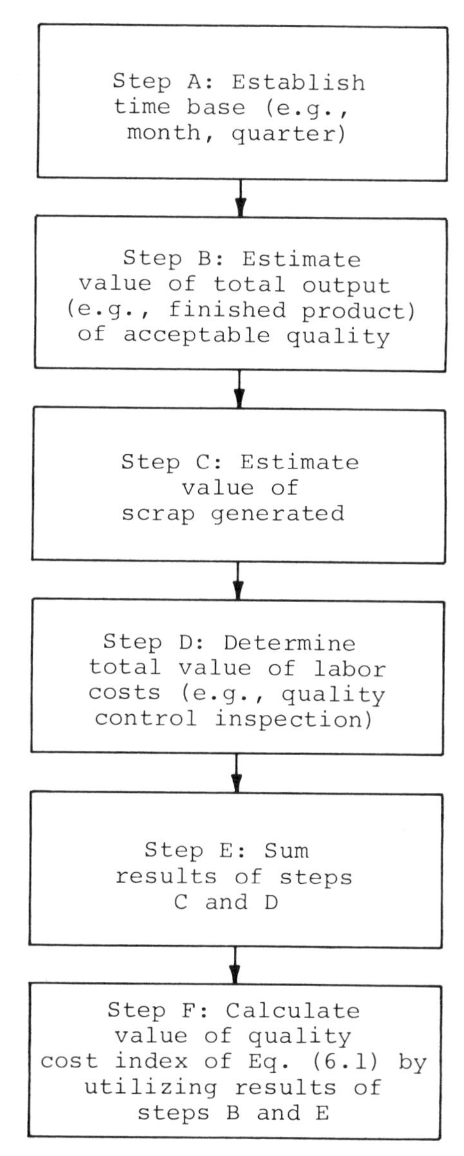

Figure 6.2 A practical approach to estimating the value of quality cost index.

and materials used has to be properly monitored: low-quality components and materials will result in poor-quality end products. Therefore, this section describes various important aspects of procurement quality control.

6.3.1 Reasons for Having Procurement Quality Assurance

There are several reasons for having procurement quality assurance. Some of them are as follows [5]:

1. To aid the suppliers of purchased items understand the specified requirements
2. To determine whether the requirements are being met
3. To ascertain whether the contractual and technical needs are adequately specified in the document known as the request for proposal
4. To rate performance of the vendor
5. To rate vendor conformance
6. To participate in the selection of vendors who meet desired qualifications

6.3.2 Areas Pertinent to a Procurement Quality Plan

Some of the areas relevant to a procurement quality plan are as follows [5]:

1. Investigation of failures
2. Source appraisal and inspection
3. Audits and rating of the performance of vendors
4. Selection and evaluation of vendors
5. Program for corrective measures
6. Plan for receiving inspection

According to Feigenbaum [6], the following items are essential to control incoming material effectively:

1. Auditing and surveying the suppliers of components and materials
2. Availability of proper material handling equipment and services
3. Availability of adequate receiving inspection and test facilities
4. Availability of adequate storage facilities
5. Availability of quality procurement information for vendors and providing other necessary information to vendors
6. Making use of statistical techniques to analyze data on incoming materials
7. Periodically reviewing the effectiveness of inspection of the incoming materials and components

8. Establishing compatible quality measurement procedures between users and suppliers
9. Maintaining measuring equipment and gauges used in the quality work
10. Establishing a close relationship with suppliers so that corrective actions can be taken
11. Providing necessary training to the inspectors of incoming material to achieve results
12. Setting up a system so that nonconforming materials and components are disposed of immediately
13. Making use of acceptance sampling tables

6.3.3 The Purchase Requisition and Specification

This section briefly discusses, separately, two essential areas related to procurement quality control.

The Purchase Requisition

This is used by the purchasing personnel to procure components, materials, and so on. Thus, its clarity plays an important role in procurement quality control. Furthermore, proper care must be taken when designing the purchase requisition—for example, when deciding the amount and type of information to be specified on the purchase requisition. The design of such requisitions may vary from one company to another; however, in any situation the purchasing requisition should include information such as authority for the requisition, the required date of the product to be purchased, quality of the product, place where the product is to be delivered, quantity of the product, product's description, account number to which its cost is to be charged, and the justification for the acquisition.

Specifications

These also play an important role in the procurement quality control. A poorly written design specification of a product may be one of the factors in poor product quality. Groups such as quality control, engineering, etc., are responsible for outlining quality requirements. However, the responsibility for procuring a product with a specified quality at the minimum cost lies with the purchasing group. In any case when writing a specification points such as the following must be considered seriously:

1. Avoid specifying conflicting requirements.
2. Avoid repeating.
3. Minimize the use of cross references.

4. Use words in their exact meanings.
5. Do not specify conditions which are impossible to meet.
6. Pay special attention so that requirements are complete and definite.
7. Do not write specifications in such a way that any difficulties are concealed from the prospective supplier.
8. Write specifications in such a way that they give directions rather than suggestions.

In a capsule form, it may be said that a specification must be accurate, complete, flexible, concise, clear, and reasonable for specified tolerances.

Specification Review Team

There are various people who form the specification review team. All these persons look into the specification from their area of concern. The size of the team depends on the complexity of the product under consideration. Usually for a fairly complex product the team is composed of the following persons [4]:

1. Tooling and design engineers
2. Representatives from quality control and purchasing groups
3. Representatives from marketing, packaging, and shipping groups
4. Engineers from manufacturing department
5. The process and field service engineers
6. Representative from manufacturing supervision

In addition, depending on the product under consideration, people such as the value engineer, safety engineer, reliability engineer, and human factor engineer also participate in specification review.

Benefits of Standard Specifications

Very many times, standard specifications are used when purchasing new materials or components. Some of the benefits of standard specifications are given below:

1. The procured items usually are less costly.
2. Standard specifications can simply be referred to.
3. They are useful in comparing bids.
4. They are useful in comparing results of tests performed by different companies.
5. They use already tried specifications.

6.3.4 Inspection

The basic objective of inspection is to determine if the procured items meet specified requirements. The inspection is performed by the buyer's personnel. Various aspects of inspection are described below.

Inspection of Incoming Materials

Many purchased items are inspected whenever they are received at the buyer's facility. The following are the some of the reasons for having the receiving inspection:

1. Products may be unsafe to workers.
2. The efficiency of manufacturing and outgoing quality will be lowered if the shipment is defective.
3. Chances are high for having defective items in the order.

The inspection of incoming materials may take various different forms. Some of them are as follows [7]:

1. The received units (i.e., purchased units) are only checked to see if they are the same as ordered.
2. All the items of the received (purchased) lot are examined for conformance to the required specifications. This type of inspection is known as a 100 percent inspection.
3. A sample of the received (purchased) lot is taken to determine if the lot conforms to the required specifications. This way the decision is made whether to accept or reject the entire received lot.

Duties of the Receiving Inspection Personnel

The receiving inspection personnel have to perform various types of duties [3]. Some of the duties performed by such personnel are as follows:

1. Accept or reject incoming materials according to specified requirements.
2. Complete necessary paperwork for accepted or rejected materials.
3. Place the accepted items at their proper locations.
4. Inspect received (purchased) items according to specified procedures.
5. Test received items (purchased) according to outlined requirements.

Source Inspection

Source inspection is concerned with inspecting the source of purchased materials or items. The degree of such surveillance depends on the product in question, the reputation of the manufacturer, and so on. Furthermore, the buyer may have a resident inspector or inspectors at the manufacturer's facility, or its inspectors may visit the manufacturer's facility occasionally. The basic functions of the buyer's quality control inspector are to make the manufacturer aware of deficiencies, monitor the adherence to buyer's requirements by the manufacturer, perform system inspections, monitor tests, and so on.

Formulas to Determine Accuracy and Waste of Inspectors

After all, inspectors are also human beings and can accept bad items and reject good ones. In this case check inspectors may review the performance of the inspectors. In other words, the check inspectors reexamine both the accepted and rejected items by the regular inspectors as well the procedure followed by them. This section presents two mathematical formulas to measure the accuracy and waste of regular inspectors. These formulas are as follows.

Formula I. From Ref. 8, θ, the percent of defects correctly identified by the regular inspector, is given by

$$\theta = \frac{\alpha}{\alpha + \beta}(100) \tag{6.2}$$

where

$$\alpha \equiv \lambda - \mu \tag{6.3}$$

The symbols used in Eq. (6.2) and (6.3) are defined as follows: β is the number of defects missed by the regular inspector as revealed by the check inspector; λ is the number of defects discovered by the regular inspector; and μ is the number of units without defects rejected by the regular inspector as revealed by the check inspector.

Formula II. This formula is used to calculate σ, the percent of good units rejected by the regular inspector. The formula is defined as follows:

$$\sigma = \frac{\mu(100)}{m - (\alpha + \beta)} \tag{6.4}$$

where m is the total number of units inspected.

Example 6.1 A regular inspector inspected a number of units in a lot and found 60 defects. All the units (i.e., good plus defective) of the lot were reexamined by the check inspector. Thus according to the findings of the check inspector the values of μ and β were 10 and 15, respectively. Calculate the percent of defects correctly discovered by the regular inspector.

By substituting the above specified data into Eqs. (6.2) and (6.3), we get

$$\theta = \frac{\alpha(100)}{\alpha + 15} \tag{6.5}$$

$$\alpha \equiv 60 - 10 = 50 \tag{6.6}$$

By utilizing the result of Eq. (6.6) in Eq. (6.5) we obtain

$$\theta = \frac{50}{50 + 15}(100) = 76.92\%$$

Thus the percent of defects correctly discovered by the regular inspector is 76.92%.

6.3.5 Rating of Vendor

To ascertain adequate quality of the procured items, the rating of vendors is necessary. Usually, the overall performance of a vendor is measured by taking into consideration factors such as delivery, quality, and cost of procured items. According to Lester et al. [4], the frequently assigned weights for delivery, quality, and cost are 30%, 30%, and 40%, respectively.

Vendor Quality Rating Formula

The quality rating Q_r is defined as follows:

$$Q_r = \frac{L_a}{L_{tr}} \times 100 \tag{6.7}$$

where L_{tr} is the total number of lots received by the buyer (i.e., lot accepted plus lot rejected), and L_a is the number of lots accepted by the buyer.

It is to be noted that the index of Eq. (6.7) does not take into consideration the delivery and cost factors.

Example 6.2 An electronic product manufacturer received 25 shipments of a certain electronic component from a vendor during the past 6-month period. Each shipment contained an equal number of components. Only three shipments were rejected out of the total shipments received. Calculate the value of the quality rating factor, Q_r.

By substituting the above data into Eq. (6.7) we get

$$Q_r = \frac{L_a}{25}(100)$$

where

$$L_a = (\text{total number of lots received}) - (\text{lots rejected})$$
$$= 25 - 3 = 22$$

Thus

$$Q_r = \frac{22}{25}(100) = 88\%$$

6.4 SUMMARY

This chapter describes, separately, various aspects associated with quality costs and procurement quality control. Thus, the chapter is divided into two parts: quality costs and procurement quality control. Five different kinds of quality costs are discussed. These include appraisal and detection costs, prevention costs, administrative costs, internal failure costs, and external failure costs. In addition, an approach to evaluating quality costs and ways in which the quality organization increases company profits are discussed. Lastly the advantages of quality costs are listed.

The other main topic discussed is procurement quality control. Reasons for having a procurement quality assurance program are given. Areas pertinent to a procurement quality plan are outlined. Essential guidelines to control incoming materials are specified. The purchase requisition and specification are described. Other areas such as specification review team, advantages of standard specifications, inspection, formulas to determine accuracy and waste of inspectors, and the vendor quality rating formula are briefly discussed.

EXERCISES

1. Describe any two of the following quality costs:
   a. Prevention costs
   b. Internal failure costs
   c. Administrative costs
2. What are the advantages of quality costs?
3. What are the reasons for having a procurement quality assurance program?

4. List at least five reasons for having inspection of incoming materials.
5. What are the responsibilities of the incoming materials inspection personnel?
6. What is the meaning of source inspection? Discuss reasons for having the source inspection.
7. What are the useful guidelines which are to be considered when writing a specification?

REFERENCES

1. Editorial, *Quality Management and Engineering Magazine*, Vol. 12, No. 3, 1972, p. 11.
2. C. L. Carter, *The Control and Assurance of Quality, Reliability and Safety*, 1978, C. L. Carter and Associates, Inc., Richardson, Texas 75080, pp. 224–225.
3. G. E. Hayes and H. G. Romig, *Modern Quality Control*, Collier-Macmillan, London, 1977, pp. 737–738.
4. R. H. Lester, N. L. Enrick, and H. E. Mottley, *Quality Control for Profit*, Industrial Press Inc., New York, pp. 254–261.
5. W. G. Gage, Procurement Quality Planning and Control, Proceedings of the Annual American Society for Quality Control Conference, 1978, pp. 158–161.
6. A. V. Feigenbaum, *Total Quality Control*, McGraw-Hill, New York, 1983, pp. 712–716.
7. J. M. Juran and F. M. Gryna, *Quality Planning and Analysis*, McGraw-Hill, New York, 1980, pp. 252–253.
8. J. M. Juran, F. M. Gryna, and R. S. Bingham, *Quality Control Handbook*, McGraw-Hill, New York, 1974, pp. 12.61–12.62.

7

Statistical Quality Control

7.1 INTRODUCTION

In statistical quality control statistics is used to attack the quality control problem. Statistical quality control has been defined as the application of statistical methods in all phases of an operation so that the established quality requirements are met at minimum cost [1].

As mentioned earlier, the origin of statistical quality control goes back to 1916 when C. N. Frazee of Bell Telephone Laboratories first made use of statistical concepts in solving inspection problems. Frazee developed the well-known operating characteristic curve. Walter A. Shewhart is another person whose contribution represents an important milestone in the history of statistical quality control. In 1924, he developed the widely known quality control charts. However, it was not until 1946 that the American Society for Quality Control (ASQC) was established. During the same year the society started the publication of a journal known as *Industrial Quality Control.* This journal was renamed *Quality Progress* in 1967. Another ASQC journal called *Journal of Quality Technology* was also started during the same year. In addition, the ASQC publishes the proceedings of its Annual Conference on Quality Control. In all these publications various researchers have published their contributions on statistical quality control. Today there are various other research publications where articles on statistical quality control are being published by researchers around the world.

This chapter briefly explores the various important aspects of statistical quality control.

7.2 REVIEW OF BASIC STATISTICS

This section presents essential basic statistics which become useful in statistical quality control.

7.2.1 Frequency Distributions

In quality control work various types of raw data are available. The useful information from such data can only be retrieved if the data are properly tabulated. Thus, frequently such data are tabulated into groups to determine the class frequency (i.e., the number of individuals belonging to each group). Such a tabular arrangement of data by groups along with their respective group frequencies is known as the *Frequency Distribution*. The following guidelines are necessary for forming frequency distributions:

1. Finding the largest and the smallest numbers in the data to calculate range (i.e., the difference between the largest and smallest numbers)
2. Apportioning the range into a desired number of groups of equal size
3. Determining the group frequencies (i.e., counting the observations of each group)

Histograms

Broadly speaking these are the graphical representations of frequency distributions and are composed of a set of rectangles. They are very useful for performing data analysis. In quality control work, the histogram can be used for various purposes, for example, comparing products, materials, suppliers, operators; determining strengths of material and capabilities of machines; and evaluating corrective measure effects and processes.

Some of the shortcomings of a simple histogram are inability to show trends, requirement for many measurements, and the inability to take into consideration the time factor.

Example 7.1 A class of 50 students was examined at the end of a three-month quality control course. Out of 100 marks, the students obtained marks as shown in Table 7.1. Construct a histogram of marks scored by students. Rearranging the marks given in Table 7.1 from smallest to largest results in Table 7.2. Grouping the data into six classes leads to the results shown in Table 7.3. A histogram of data given in Table 7.3 is shown in Fig. 7.1.

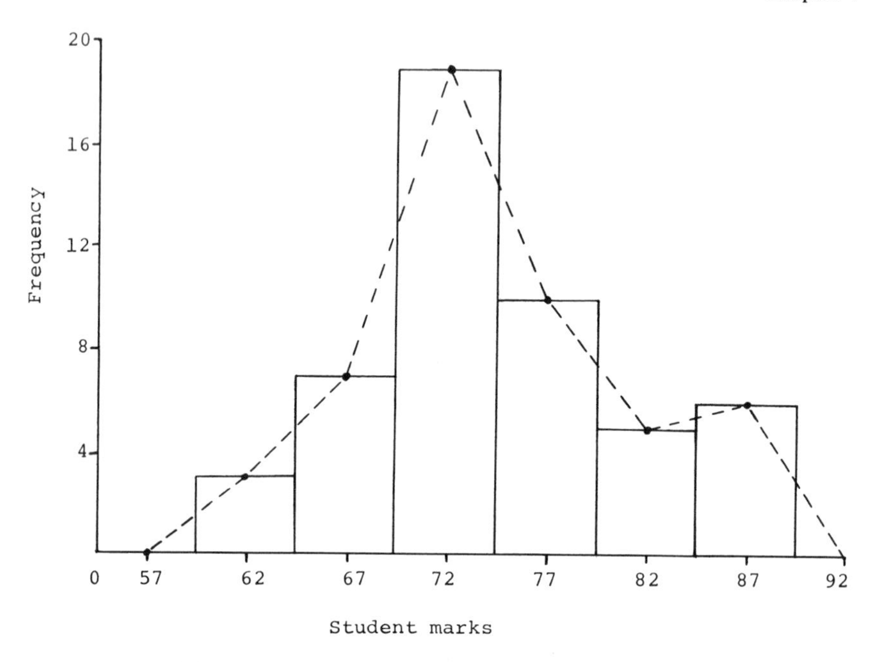

Figure 7.1 Histogram of data given in Table 7.3.

Table 7.1 Marks Scored by Students

60	89	76	89	70
65	70	75	75	82
67	73	85	73	79
68	69	65	74	72
70	70	70	70	74
85	75	64	83	66
89	77	83	69	81
80	78	74	63	73
75	85	73	74	77
73	75	74	72	72

Table 7.2 Rearranged Marks

60	70	73	75	81
63	70	73	75	82
64	70	73	75	83
65	70	73	75	83
65	70	74	76	85
66	70	74	77	85
67	72	74	77	85
68	72	74	78	89
69	72	74	79	89
69	73	75	80	89

Table 7.3 Marks of Fifty Students

Class	Class midpoint	Frequency
60–64	62	3
65–69	67	7
70–74	72	19
75–79	77	10
80–84	82	5
85–89	87	6

7.2.2 The Arithmetic Mean

This is a widely used measure of location and is given by

$$\bar{N} = \frac{\sum_{i=1}^{k} n_i}{k} \tag{7.1}$$

where n_i is the ith number, k is the total number of numbers in a set, and $\bar{N}$ is the arithmetic mean of numbers in a set.

Example 7.2 Calculate the arithmetic mean of the following set of 10 numbers:

[20, 15, 25, 30, 40, 35, 60, 75, 100, 80]

In this example, the values of k and n_i are as follows:

$k = 10, \quad n_1 = 20, \quad n_2 = 15, \quad n_3 = 25, \quad n_4 = 30, \quad n_5 = 40, \quad n_6 = 35$
$n_7 = 60, \quad n_8 = 75, \quad n_9 = 100, \quad n_{10} = 80$

Thus substituting the above values into Eq. (7.1) results in

$$\bar{N} = \frac{(20) + (15) + (25) + (30) + (40) + (35) + (60) + (75) + (100) + (80)}{10}$$
$$= 48$$

The arithmetic mean of the set of numbers is 48.

7.2.3 The Range

This is given by the difference between the largest and smallest numbers contained in a set. Thus,

$$r = N_h - N_\ell \tag{7.2}$$

where r is the range, N_h is the highest number in the set, and N_ℓ is the lowest number in the set.

Example 7.3 Compute the range of the following set of numbers:

[10, 4, 108, 10, 40, 120, 80, 100, 70, 95]

Utilizing the above data in Eq. (7.2) results in

$$r = 120 - 4 = 116$$

The range of the set of numbers is 116.

7.2.4 The Mean Deviation

This is used as a measure of variability. In other words it is the mean deviation of a set of numbers from their mean value and is given by

$$D_m = \frac{\sum_{i=1}^{k} |n_i - \bar{N}|}{k} \tag{7.3}$$

where

D_m = the mean deviation

n_i = the ith number of observation
k = the number of observations or the total number of numbers in a set
$\bar{N}$ = the arithmetic mean
$|n_i - \bar{N}|$ = the absolute deviation of n_i from $\bar{N}$; in other words, the absolute value does not take into consideration the signs (all deviations are considered as positive)

Example 7.4 Times to failure of seven identical mechanical components are 200, 205, 196, 200, 210, 180, and 209 hr. Calculate the value of the mean deviation.

Substituting the above data into Eq. (7.1) leads to

$$\bar{N} = \frac{(200) + (205) + (196) + (200) + (210) + (180) + (209)}{7}$$

$$= 200 \text{ hr}$$

Utilizing the above data in Eq. (7.3) results in

$$D_m = \frac{|200 - 200| + |205 - 200| + |196 - 200| + |200 - 200| + |210 - 200| + |180 - 200| + |209 - 200|}{7}$$

$$= \frac{|0| + |5| + |-4| + |0| + |10| + |-20| + |9|}{7}$$

$$= \frac{0 + 5 + 4 + 0 + 10 + 20 + 9}{7} = 6.857$$

The mean deviation is 6.857.

7.2.5 The Standard Deviation

Sometimes, standard deviation is also known as the root mean square deviation. It is simply a measure of the dispersion. The standard deviation σ of a group of numbers is given by

$$\sigma = \left[\frac{\sum_{i=1}^{k} (n_i - \bar{N})^2}{k} \right]^{1/2} \tag{7.4}$$

where σ is the root mean square of the deviations from the mean value.

Other symbols used in Eq. (7.4) are defined in the earlier section of the chapter. The standard deviation possesses certain properties. For example, in the case of normal distributions, it has the following properties:

1. 99.73% of the values are included within $\bar{N} \pm 3\sigma$.
2. 95.45% of the values are included within $\bar{N} \pm 2\sigma$.
3. 68.26% of the values are included within $\bar{N} \pm \sigma$.

Example 7.5 Determine the standard deviation of data given in Example 7.4.

From Example 7.4, the arithmetic mean is 200 hr. Using the specified data in Eq. (7.4) results in

$$\sigma = \left[\frac{(200-200)^2 + (205-200)^2 + (196-200)^2 + (200-200)^2 + (210-200)^2 + (180-200)^2 + (209-200)^2}{7} \right]^2$$

$$= \left[\frac{(0)^2 + (5)^2 + (-4)^2 + (0)^2 + (10)^2 + (-20)^2 + (9)^2}{7} \right]^{1/2}$$

$$= 99.426$$

Thus the standard deviation of the data set is 9.426.

7.2.6 The Variance

The variance is given by the square of the standard deviation of a set data. Thus

$$\text{Variance} = (\text{standard deviation})^2 = \sigma^2$$

$$= \frac{\sum_{i=1}^{k} (n_i - \bar{N})^2}{k} \tag{7.5}$$

Example 7.6 Times to failure of five identical electronic components are 500, 495, 510, 490, and 520 hr. Determine the variance of the data set.

By substituting the specified data into Eq. (7.1), we get the arithmetic mean

$$\bar{N} = \frac{(500) + (495) + (510) + (490) + (520)}{5}$$

$$= 503 \text{ hr}$$

Again using the specified data and the above result in Eq. (7.5) leads to

$$\text{Variance} = \frac{(500-503)^2 + (495-503)^2 + (510-503)^2 + (490-503)^2 + (520-503)^2}{5}$$

$$= \frac{(-3)^2 + (-8)^2 + (7)^2 + (13)^2 + (17)^2}{5} = 116$$

Thus the variance of the data set is 116.

7.3 ACCEPTANCE SAMPLING

This is probably the oldest technique for judging quality. It is a process for making statistical decisions on the quality of items contained in a lot. Economy is the important advantage of sampling because it helps to reduce the cost of inspection. For example, when a large number of items are produced, the sampling will reduce the cost of inspection significantly because fewer items will be inspected. The sampling concept is based on the assumption that the sample taken from a population implies characteristics of that population. For example, it is assumed that the percentage of bad items found in a sample will be the same as that of the population.

In addition, in many cases it is physically impossible to inspect all the units of a certain product. Some examples of such cases are tasting each piece of candy before packing, firing each rocket to examine whether it functions normally, firing each piece of ammunition, testing each straight pin for sharpness, and inspecting each foot of coils of cable.

This section describes the various aspects of acceptance sampling.

7.3.1 Benefits and Drawbacks of Sampling

The following are some of the benefits and drawbacks of sampling [2].

Benefits

The benefits of sampling are as follows:

1. The handling damage during inspection is less.
2. It provides a stronger motivation for improvement because entire lots are rejected instead of returning only the defective items.
3. Sampling is applicable to destructive testing.
4. It is less costly because only a small portion of the total items are inspected.

5. It helps to simplify the recruiting and training problem because of the need for fewer inspectors.
6. There are fewer inspection errors in comparison to 100 percent inspection.
7. The problem of monotony is reduced.

Drawbacks

The drawbacks of sampling are as follows:

1. There are certain sampling risks, for example, rejecting the lots containing good items and accepting the ones having bad units.
2. There is more administrative work and costs.
3. It provides less information on the product in question as compared with 100 percent inspection.

7.3.2 Lot Formation and Sampling Error

This section briefly describes the lot formation and sampling error. A lot is a collection of similar items or materials from a common source. Thus, careful consideration must be given when forming a lot. The outgoing quality of the product will be strongly influenced by how the lots were formed. According to Ref. 3, the following guidelines must be taken into consideration when forming a lot:

1. Try to make the size of the lot as large as practicable, since usually the inspection cost decreases as the lot size becomes larger.
2. Avoid accumulating manufactured goods over a long period.
3. Make sure that the lots conform to healthy packaging practices.
4. Utilize extraneous information on items such as process capability and prior inspections when forming a lot.
5. Make certain that lots are homogeneous.
6. Make certain that the lots are designed so that they satisfy the material handling requirements.

The sample size plays an important role in the occurrence of sampling error. For example, the sampling risk reduces as the size of the sample increases. Therefore, one has to choose a sample and its size in such a way that the risk of sampling is minimum and cost effective. The mathematics is used to estimate risk of sampling. The usual procedure is to establish upper and lower accuracy limits. Thus from Ref. 4, the following formula can be used to compute upper and lower accuracy limits (the formula is associated with the binomial distribution):

$$F \pm 3 \left[\frac{F(1-F)}{k} \right]^{1/2} \tag{7.6}$$

where F is the percentage of defective items in a sample and k is the sample size.

Thus from Eq. (7.6), the upper and lower accuracy limits, respectively, are as follows:

$$\text{Upper limit} = F + 3 \left[\frac{F(1-F)}{k} \right]^{1/2} \tag{7.7}$$

and

$$\text{Lower limit} = F - 3 \left[\frac{F(1-F)}{k} \right]^{1/2} \tag{7.8}$$

The single sample will show F percent defective, but the sampling error limits will be as given by Eqs. (7.7) and (7.8).

Example 7.7 A lot contains 10,000 mechanical components. A sample of 1,000 components was selected from that lot. During the examination of the sample, it was found that the sample contains 90 defective units. Determine the upper and lower limits of sampling error.

By utilizing the specified data we get the percentage of defectives in the sample as follows:

$$F = \frac{\text{number of defects} \times (100)}{\text{number of units in the sample}} = \frac{90 \times (100)}{1,000} = 9\%$$

Substituting the above data into Eq. (7.7) and (7.8) results in

$$\text{Upper limit} = 0.09 + 3 \left[\frac{0.09(1-0.09)}{1,000} \right]^{1/2} = 11.72\%$$

and

$$\text{Lower limit} = 0.09 - 3 \left[\frac{0.09(1-0.09)}{1,000} \right]^{1/2} = 6.29\%$$

thus the upper and lower limits of sampling error are 11.72% and 6.29%, respectively.

7.3.3 Sampling Plans

There are three types of such plans which are used in lot-by-lot inspection. These are

1. Single sampling
2. Double sampling
3. Multiple sampling

The above three plans are described below, separately.

Single Sampling Plan

In the single sampling case, the lot is accepted or rejected on the basis of the result of inspecting a single sample drawn from that lot. Schematically, the single sampling plan is described in Fig. 7.2.

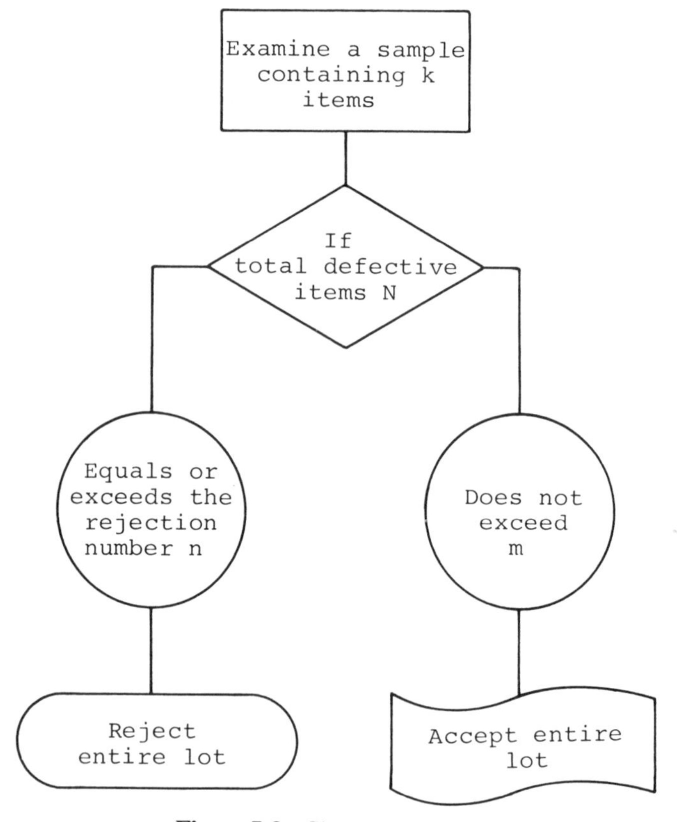

Figure 7.2 Single sampling plan.

In Fig. 7.2 m denotes the allowable defectives in a sample. Sometimes it is known as the acceptance number. In addition, n denotes the rejection number. Both acceptance and rejection numbers are tabulated in MIL-STD-105 D and in Ref. 3 and 4.

Double Sampling Plan

The double sampling plan is very much same as the single sampling plan [4]. However, this plan calls for two samples, two acceptance numbers, and two rejection numbers. In the double sampling plan usually the size of the first sample is smaller than the second sample drawn from the same lot. In the case of the first sample the decision to accept the lot is only taken if the total defectives found in the sample are equal to or less than the first acceptance number. However, on the other hand, the entire lot is rejected if the total defectives found in the first sample are equal to or exceed the first rejection number in the sampling plan. The second sample is only drawn if the defectives of the first sample fall between the first acceptance and rejection numbers. Furthermore, the entire lot is only accepted if the combined defectives found in both samples (i.e., the defectives found in both the first and second samples) are less than or equal to the second acceptance number. Similarly, the lot is rejected if the combined number of defectives found in both samples (i.e., first and second samples) is equal to or exceeds the rejection number associated with the second sample. Schematically, the double sampling plan is described in Fig. 7.3.

In Fig. 7.3, m_i denotes the ith acceptance number or allowable defectives in a sample, for $i = 1$ (first sample) and $i = 2$ (second sample).

Multiple Sampling Plan

Basically, the multiple sampling plan is a continued version of the double sampling plan. For example, after examining the second sample, if the decision cannot be reached whether to accept or reject the lot, then another sample (i.e., the third sample) is taken and inspected. Again if the decision cannot be made, then the fourth sample is examined. In this way the process continues. Usually, according to Ref. 4, the MIL-STD-105 D allows a maximum of seven samples. Schematically, the multiple sampling plan is described in Fig. 7.4. In Fig. 7.4, m_i denotes the ith acceptance number or allowable defectives in a sample, for $i = 1$ (first sample), $i = 2$ (second sample), and $i = 3$ (third sample).

7.3.4 The Operating Characteristic Curve

According to Ref. 4, C. N. Frazee of Telephone Laboratories first made use of the operating characteristic (OC) curve. The OC curve is a curve depicting

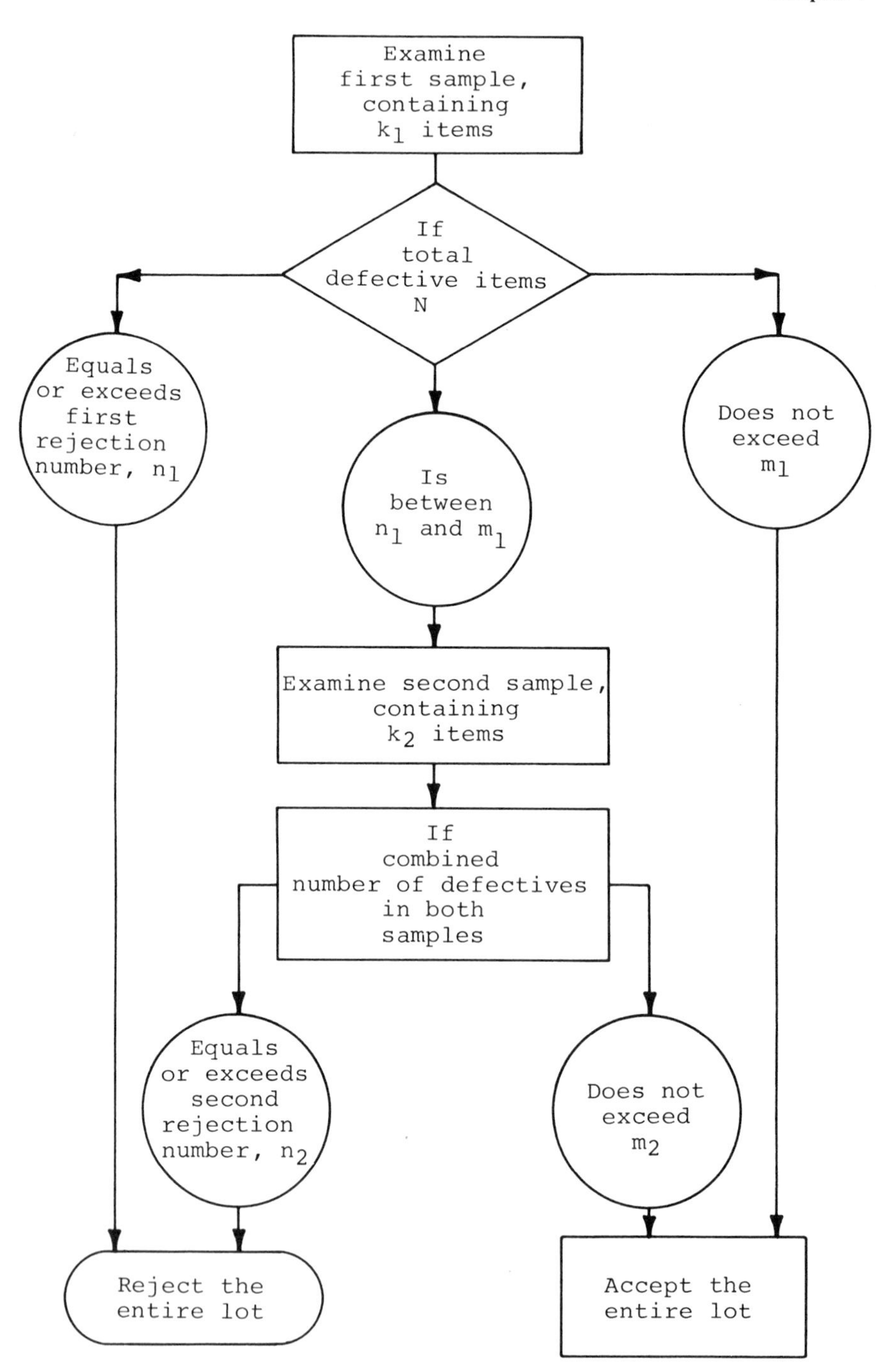

Figure 7.3 Schematic description of the double sampling plan.

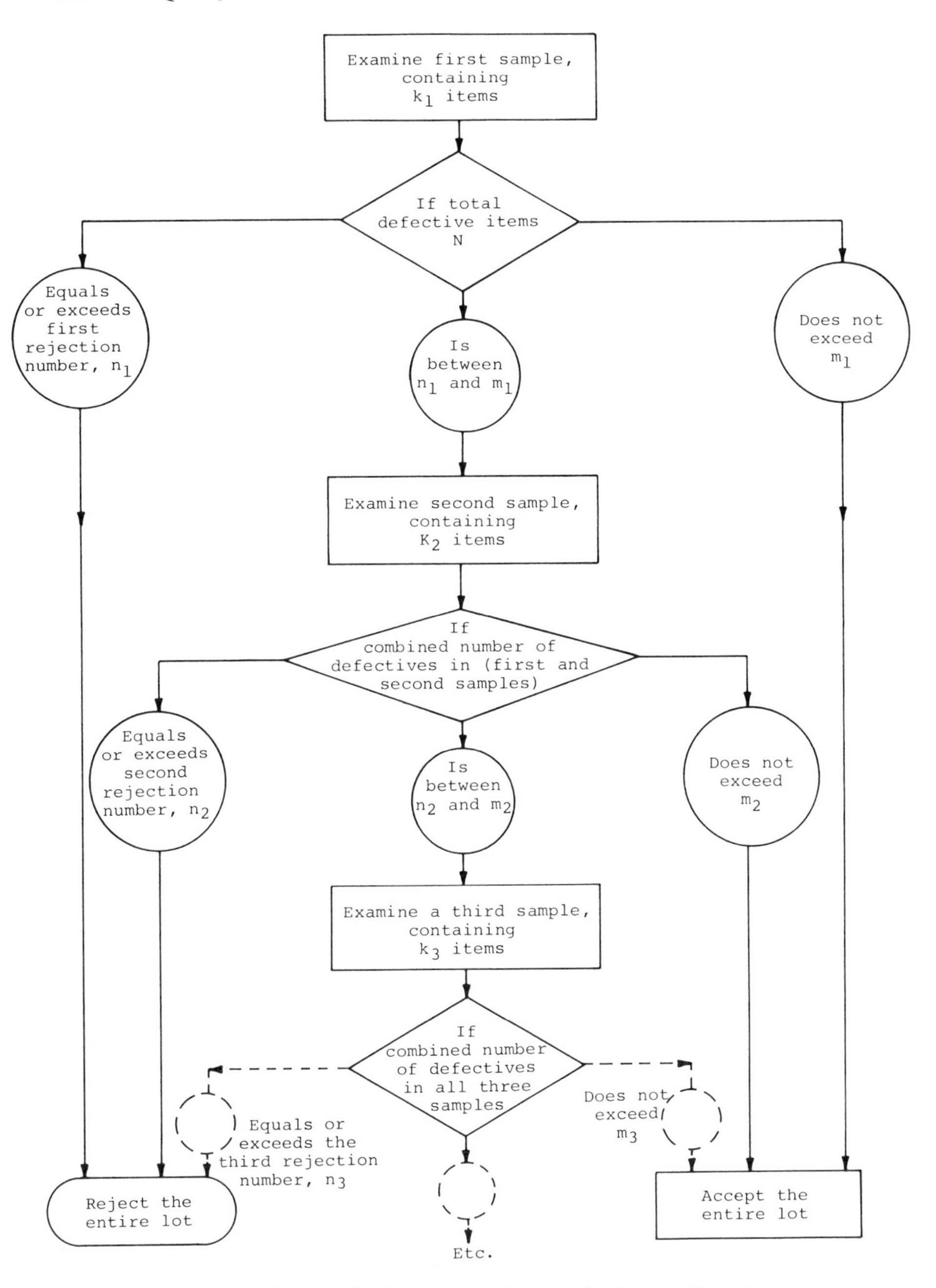

Figure 7.4 Schematic description of the multiple sampling plan.

for a specified sampling plan the acceptance probability of a lot. Neither 100 percent inspection nor sampling can assure that all the defectives in a lot will be discovered. Both procedures have associated risks. For example, in sampling there is a risk of accepting the bad lot and rejecting the good one. Similarly in 100 percent inspection there is a certain degree of risk that inspectors will miss some defectives due to monotony and other factors.

The following four parameters are concerned with a OC curve:

1. The producer's risk parameter: This is usually denoted by α. The parameter α represents the probability that the sampling plan will reject a good lot of items.
2. The consumer's risk parameter: This is usually denoted by β. The parameter β represents the probability that the sampling plan will accept a bad lot of items.
3. The lot tolerance percent defective (LTPD) parameter: This is denoted by LTPD. The parameter LTPD represents that percent defective in a lot of items is at unacceptable quality level, at which the acceptance probability is very low, usually at 10% or less. It is associated with the consumer's risk.
4. The acceptance quality level (AQL) parameter: This is denoted by AQL. The parameter AQL represents the maximum percentage of defective items, for the acceptance sampling purpose, that can be considered satisfactory as a process average [2].

An ideal operating characteristic (OC) curve is shown in Fig. 7.5. Figure 7.6 shows a typical operating characteristic curve with α, β, LTPD, and AQL.

The OC curve in Fig. 7.5 depicts that the lots containing more than 6% defective units would be rejected and the lots containing less than 6% defective units would be accepted. The OC curve in Fig. 7.5 can only be achieved through 100 percent inspection. Even 100 percent inspection has its shortcomings. Therefore, it is safe to say that there is no sampling plan which can discriminate perfectly.

The Operating Characteristic Curve Construction

The Poisson distribution is used to construct the operating characteristic curve. Thus the probability P_k of having exactly k defective units in sample of m units is given by

$$P_k = \frac{e^{-mq}(mq)^k}{k!} \tag{7.9}$$

where q is the percentage of defective units in a lot.

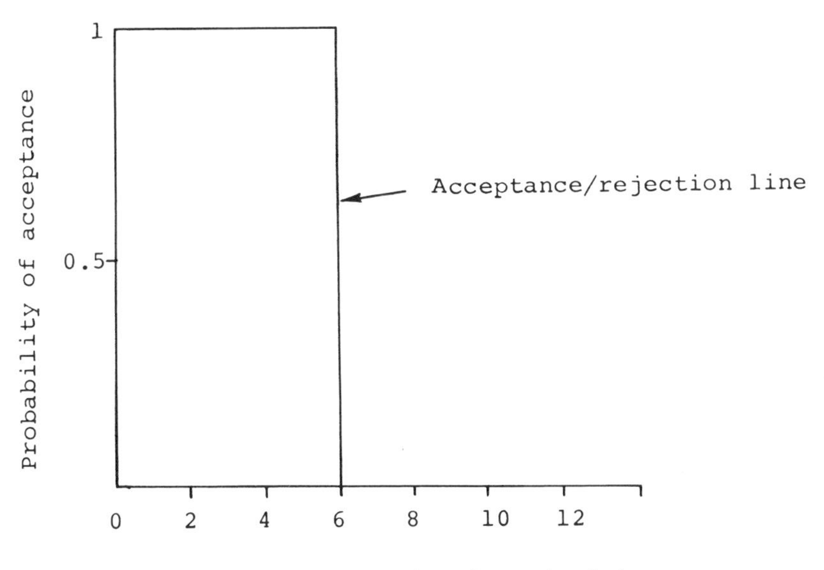

Figure 7.5 The operating characteristic curve of a perfect sampling plan.

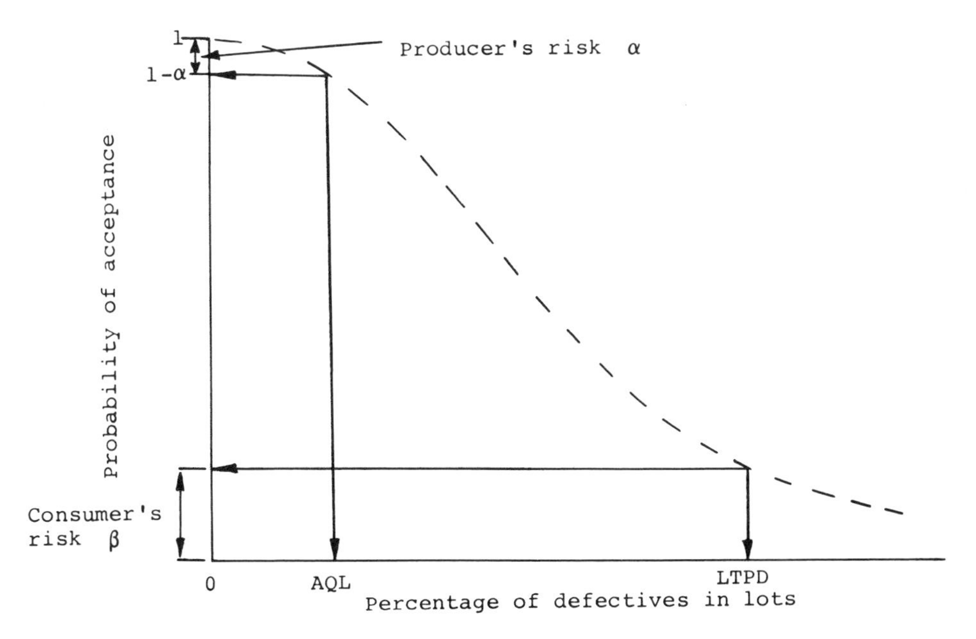

Figure 7.6 A typical operating characteristic curve.

Example 7.8 A lot contains 200 components. A sample of 30 components is to be selected from the lot. Develop an operating characteristic curve if the percentage of defectives in the lot varies from 0 to 13 percent. In addition, the lot should only be accepted if 1 or fewer defective components are discovered in the sample.

In the above example the data are specified for the following symbols of Eq. (7.9):

$$m = 30,\quad q_1 = 0,\quad q_2 = 1\%,\quad q_3 = 2\%,\quad q_4 = 3\%,\quad q_5 = 4\%,$$
$$q_6 = 5\%,\quad q_7 = 6\%,\quad q_8 = 7\%,\quad q_9 = 8\%,\quad q_{10} = 9\%,$$
$$q_{11} = 10\%,\quad q_{12} = 11\%,\quad q_{13} = 12\%,\quad q_{14} = 13\%.$$

For $m = 30$, $q_1 = 0$. From Eq. (7.9) we get

$$P_0 = [(30)(0)]^0 e^{-(30)(0)}/0!$$
$$= 1$$

and

$$P_1 = [(30)(0)]^1 e^{-(30)(0)}/1!$$
$$= 0$$

Where P_0 is the probability of having exactly zero defective in a sample of 30 components and P_1 is the probability of having exactly one defective in a sample of 30 components. Thus the probability of accepting the lot P_a is

$$P_a = P_0 + P_1 = 1 + 0 = 1 \tag{7.10}$$

For $m = 30$, $q_2 = 0.01$. From Eq. (7.9) we get

$$P_0 = [(30)(0.01)]^0 e^{-(30)(0.01)}/0!$$
$$= 0.7408$$

and

$$P_1 = [(30)(0.01)] e^{-(30)(0.01)}/1!$$
$$= 0.2222$$

Thus the probability of accepting the lot is

$$P_a = P_0 + P_1 = 0.7408 + 0.2222 = 0.9630 \tag{7.11}$$

For $m = 30$, $q_3 = 0.02$. From Eq. (7.9) we get

$$P_0 = [(30)(0.02)]^0 e^{-(30)(0.02)}/0!$$
$$= 0.5488$$

and

$$P_1 = [(30)(0.02)]^1 e^{-(30)(0.02)}/1!$$
$$= 0.3293$$

Thus the probability of accepting the lot is

$$P_a = P_0 + P_1 = 0.5488 + 0.3293 = 0.8781 \qquad (7.12)$$

For $m = 30$, $q_4 = 0.03$. From Eq. (7.9) we get

$$P_0 = [(30)(0.03)]^0 e^{-(30)(0.03)}/0!$$
$$= 0.4066$$

and

$$P_1 = [(30)(0.03)]^1 e^{-(30)(0.03)}/1!$$
$$= 0.3659$$

Thus the probability of accepting the lot is

$$P_a = P_0 + P_1 = 0.4066 + 0.3659 = 0.7725 \qquad (7.13)$$

For $m = 30$, $q_5 = 0.04$. From Eq. (7.9) we get

$$P_0 = [(30)(0.04)]^0 e^{-(30)(0.04)}/0!$$
$$= 0.3012$$

and

$$P_1 = [(30)(0.04)]^1 e^{-(30)(0.04)}/1!$$
$$= 0.3614$$

Thus the probability of accepting the lot is

$$P_a = P_0 + P_1 = 0.3012 + 0.3614 = 0.6626 \qquad (7.14)$$

For $m = 30$, $q_6 = 0.05$. From Eq. (7.9) we get

$$P_0 = [(30)(0.05)]^0 e^{-(30)(0.05)}/0!$$
$$= 0.2231$$

and

$$P_1 = [(30)(0.05)]^0 e^{-(30)(0.05)}/1!$$
$$= 0.3347$$

Thus the probability of accepting the lot is

$$P_a = P_0 + P_1 = 0.2231 + 0.3347 = 0.5578 \qquad (7.15)$$

For $m = 30$, $q_7 = 0.06$. From Eq. (7.9) we get

$$P_0 = [(30)(0.06)]^0 e^{-(30)(0.06)}/0! = 0.1653$$

and

$$P_1 = [(30)(0.06)]^1 e^{-(30)(0.06)}/1! = 0.2975$$

Thus the probability of accepting the lot is

$$P_a = P_0 + P_1 = 0.1653 + 0.2975 = 0.4628 \quad (7.16)$$

For $m = 30$, $q_8 = 0.07$. From Eq. (7.9) we get

$$P_0 = [(30)(0.07)]^0 e^{-(30)(0.07)}/0! = 0.1225$$

and

$$P_1 = [(30)(0.07)]^1 e^{-(30)(0.07)}/1! = 0.2572$$

Thus the probability of accepting the lot is

$$P_a = P_0 + P_1 = 0.1225 + 0.2572 = 0.3797 \quad (7.17)$$

For $m = 30$, $q_9 = 0.08$. From Eq. (7.9) we get

$$P_0 = [(30)(0.08)]^0 e^{-(30)(0.08)}/0! = 0.0907$$

and

$$P_1 = [(30)(0.08)]^1 e^{-(30)(0.08)}/1! = 0.2177$$

Thus the probability of accepting the lot is

$$P_a = P_0 + P_1 = 0.0907 + 0.2177 = 0.3084 \quad (7.18)$$

For $m = 30$, $q_{10} = 0.09$. From Eq. (7.9) we get

$$\mathrm{P}_0 = [(30)(0.09)]^0 e^{-(30)(0.09)}/0! = 0.0672$$

and

$$P_1 = [(30)(0.09)]^1 e^{-(30)(0.09)}/1! = 0.1815$$

Thus the probability of accepting the lot is

$$P_a = P_0 + P_1 = 0.0672 + 0.1815 = 0.2487 \tag{7.19}$$

For $m = 30$, $q_{11} = 0.10$. From Eq. (7.9) we get

$$\begin{aligned} P_0 &= [(30)(0.10)]^0 e^{-(30)(0.10)}/0! \\ &= 0.05 \end{aligned}$$

and

$$\begin{aligned} P_1 &= [(30)(0.10)]^1 e^{-(30)(0.10)}/1! \\ &= 0.15 \end{aligned}$$

Thus, the probability of accepting the lot is

$$P_a = P_0 + P_1 = 0.5 + 0.15 = 0.20 \tag{7.20}$$

For $m = 30$, $q_{12} = 0.11$. From Eq. (7.9) we get

$$\begin{aligned} P_0 &= [(30)(0.11)]^0 e^{-(30)(0.11)}/0! \\ &= 0.0369 \end{aligned}$$

and

$$\begin{aligned} P_1 &= [(30)(0.11)]^1 e^{-(30)(0.11)}/1! \\ &= 0.1217 \end{aligned}$$

Thus the probability of accepting the lot is

$$P_a = P_0 + P_1 = 0.0369 + 0.1217 = 0.1586 \tag{7.21}$$

For $m = 30$, $q_{13} = 0.12$. From Eq. (7.9) we get

$$\begin{aligned} P_0 &= [(30)(0.12)]^0 e^{-(30)(0.12)}/0! \\ &= 0.027 \end{aligned}$$

and

$$\begin{aligned} P_1 &= [(30)(0.12)]^1 e^{-(30)(0.12)}/1! \\ &= 0.098 \end{aligned}$$

Thus the probability of accepting the lot is

$$P_a = P_0 + P_1 = 0.027 + 0.098 = 0.1254 \tag{7.22}$$

For $m = 30$, $q_{14} = 0.13$. From Eq. (7.9) we get

$$\begin{aligned} P_0 &= [(30)(0.13)]^0 e^{-(30)(0.13)}/0! \\ &= 0.0202 \end{aligned}$$

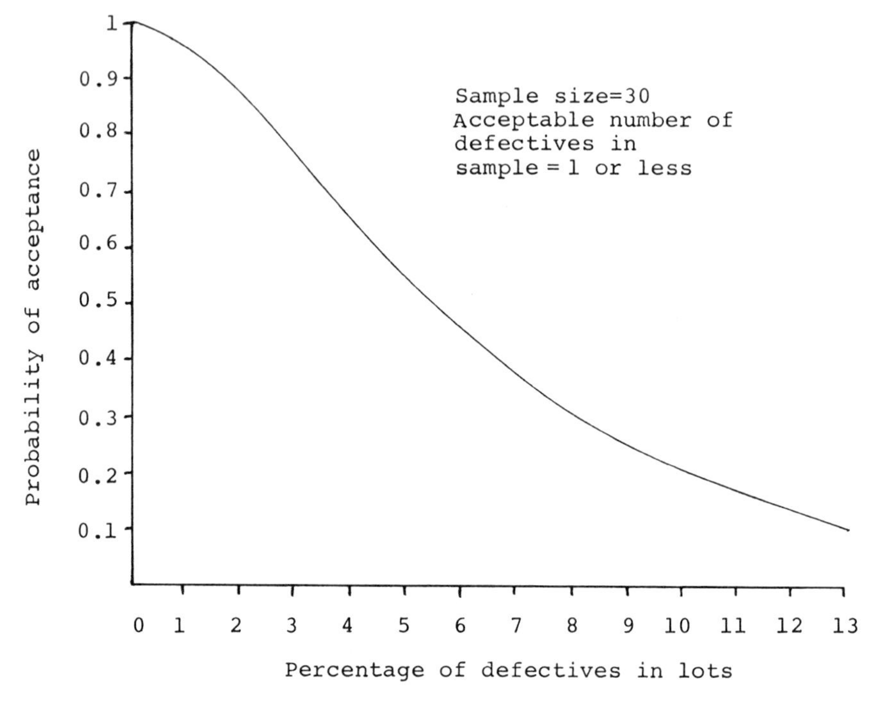

Figure 7.7 The operating characteristic curve for Example 7.8.

and

$$
\begin{aligned}
P_1 &= [(30)(0.13)]^1 e^{-(30)(0.13)}/1! \\
&= 0.079
\end{aligned}
$$

Thus the probability of accepting the lot is

$$P_a = P_0 + P_1 = 0.02 + 0.079 = 0.099 \qquad (7.23)$$

Figure 7.7 shows the plot of acceptance probabilities of Eqs. (7.10)–(7.23) with respect to their corresponding values q_i, for $i \doteq 1, 2, 3, \ldots,$ 14.

Properties of OC Curve

From Refs. 2 and 3 the properties of the OC curve are as follows:

1. The operating characteristic curves are very similar for constant sample size.

2. The operating characteristic curve becomes steeper for decreasing values of the acceptance number (i.e., acceptable number of defectives in sample).
3. The operating characteristic curve becomes steeper for increasing values of the sample size.
4. The operating characteristic curve becomes concave upward at the zero value of the acceptance number.
5. The closest method to the ideal OC curve is to increase the values of the acceptance number and sample size simultaneously.

Determining Acceptance Number and Sample Size for Specified Values of α, β, AQL, and LTPD

When the values of α, β, AQL, and LTPD are specified, the size of the sample and the value of the acceptance number can be estimated with the aid of cumulative probability distribution curves of the Poisson distribution (i.e., the Thorndike chart)—in other words, those values of both these items which will satisfy the specified operating characteristic curve. A solved numerical example is given in Ref. 5.

7.4 QUALITY CONTROL CHARTS

Control charts were developed by Walter A. Shewhart in 1924. Today these charts are used to serve various purposes: for example, to obtain information whether the process is in the state control or not, to provide information for decisions concerning inspection procedures or product specifications, and so on. According to Ref. 5, the control chart is a graphical technique used to evaluate whether a process is in a "state of statistical control" or out of control. In other words, when a sample value falls outside the upper and control limits, it signifies that the process is out of statistical control. Therefore it indicates the room for investigation to find the cause of its being out of control and then taking necessary measures. The following four types of quality control charts are described in the chapter:

1. The p-chart
2. The $\bar{X}$-chart
3. The R-chart
4. The c-chart

7.4.1 The p-Charts

These are also known as the control charts for attributes. In this case the population is classified into two categories: for example, the components with defects versus components free of defects. Thus in this situation, the binomial distribution is used to establish the control chart limits. The upper and lower control limits are given by

$$\mathrm{UCL}_p = \mu + 3\sigma \tag{7.24}$$

and

$$\mathrm{LCL}_p = \mu - 3\sigma \tag{7.25}$$

where

UCL_p = the upper control limit of the p-chart
LCL_p = the lower control limit of the p-chart
μ = the mean of the binomial distribution
σ = the standard deviation of the binomial distribution

The mean μ for the binomial distribution is given by

$$\mu = \frac{N}{m\beta} \tag{7.26}$$

where N is the total number of defectives in classification, β is the number of samples, and, m is the size of the sample.

The standard deviation σ for the binomial distribution is given by

$$\sigma = [\mu(1 - \mu)/m]^{1/2} \tag{7.27}$$

Example 7.9 Eight samples were taken from a production line. Each sample consists of 40 mechanical components. After inspection it was found that samples 1, 2, 3, 4, 5, 6, 7, and 8 contain 5, 6, 4, 2, 8, 10, 12, and 9 defectives, respectively. Develop the p-chart.

The fraction of defectives in sample 1 is

$$\frac{\text{Total number of defectives in sample 1}}{\text{Total number of items in sample 1}} = \frac{5}{40} = 0.1250$$

Similarly the fraction of defectives in samples 2, 3, 4, 5, 6, 7, and 8 are 0.15, 0.10, 0.05, 0.20, 0.25, 0.3, and 0.225, respectively.

Substituting the given data into Eq. (7.26) results in

$$\mu = \frac{56}{(40)(8)} = 0.175$$

Similarly substituting the data into Eq. (7.27) leads to

$$\sigma = \left[\frac{0.175(1 - 0.175)}{40}\right]^{1/2} = 0.060$$

Thus utilizing the calculated values for μ and σ in Eqs. (7.24) and (7.25), we get

$$UCL_p = \mu + 3\sigma = 0.175 + 3(0.06) = 0.3552$$

and

$$LCL_p = \mu - 3\sigma = 0.175 - (3)(0.06) = 0$$

For the calculated values a p-chart is shown in Fig. 7.8. The dots in the diagram signify the fraction of defectives in each sample. All the sample fractions are within the control limits, which means that there is no abnormality.

7.4.2 The $\bar{X}$-Charts

These are also known as control charts for averages. These charts belong to the family of control charts for variables and are more sensitive and effective than the control charts for attributes. The p-charts belong to the family of control charts for attributes. The $\bar{X}$-charts are used to indicate the deviation in the process mean value. The mean value of each sample is plotted on the chart.

The five-step procedure shown in Fig. 7.9 is practiced to develop both $\bar{X}$-chart and range (R) chart. According to Ref. 3, the following guidelines are helpful when selecting the sample size:

1. The distribution of sample averages, $\bar{X}$'s, follows approximately the normal distribution when the size of each sample is four or more, in other words, when each sample contains four or more items or components.
2. The sample size of five is often used in industry because of simplicity in analysis.
3. The cost of inspection per sample increases with the increase in the size of the sample.
4. The control chart becomes more sensitive to minute variations in the process mean value as the size of sample increases. In addition, the upper and lower control limits become closer to the process mean value.

The mean of the sample means is given by

$$\bar{\bar{x}} = \frac{\sum_{i=1}^{k} \bar{x}_i}{k} \tag{7.28}$$

where k denotes the total number of samples and x_i denotes the average value of the ith sample.

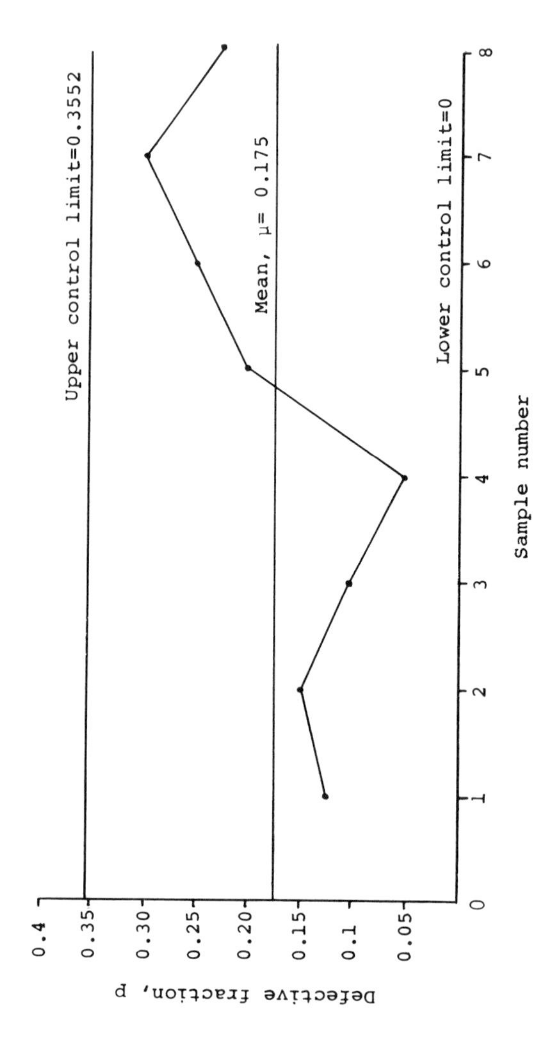

Figure 7.8 p-Chart for mechanical components.

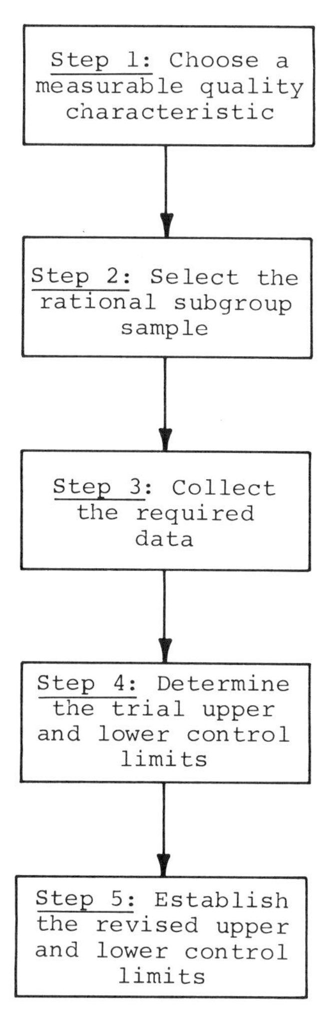

Figure 7.9 A procedure to develop $\bar{X}$ and R (range) charts.

Similarly the upper and lower control limits, respectively, are defined as follows:

$$\mathrm{UCL}_{\bar{x}} = \bar{\bar{X}} + 3s \tag{7.29}$$

and

$$\mathrm{UCL}_{\bar{x}} = \bar{\bar{X}} - 3s \tag{7.30}$$

where s is the sample averages ($\bar{x}$'s) standard deviation, $\mathrm{UCL}_{\bar{x}}$ is the upper control limit of the $\bar{X}$-chart, and $\mathrm{LCL}_{\bar{x}}$ is the lower control limit of the $\bar{X}$-chart.

In real life, to simplify computation, the $\mathrm{UCL}_{\bar{x}}$ and $\mathrm{LCL}_{\bar{x}}$ are defined as follows:

$$\mathrm{UCL}_{\bar{x}} = \bar{\bar{X}} + A_2\bar{r} \tag{7.31}$$

and

$$\mathrm{LCL}_{\bar{x}} = \bar{\bar{X}} - A_2\bar{r} \tag{7.32}$$

where

$$A_2 = \frac{3}{m^{1/2} d_2} \tag{7.33}$$

m is the size of the sample, d_2 is the factor for the sample size, and $\bar{r}$ is the mean value of the sample ranges. (The sample range is given by the difference between the highest value and the lowest value in the sample.)

For different sample sizes the values of d_2 are tabulated in Table 7.4.

Example 7.10 On an assembly line a task is performed repetitively. Over a period of time the time data were collected randomly ten times to perform such a task. Each time, the task was observed seven times and its corresponding times are given in Table 7.5. In other words, each of the ten samples contains seven readings. Develop the $\bar{X}$-chart.

The ranges of samples A to J are tabulated in Table 7.6. The sample range is given by the difference between the highest observation time and the lowest observation in the sample. This way the range for each sample was calculated. Table 7.6 shows such ranges.

From Table 7.6 the mean value of the ranges is

$$\bar{r} = \frac{12 + 11 + 12 + 7 + 13 + 11 + 7 + 9 + 12 + 9}{10} = 10.3$$

where $\bar{r}$ is the mean value of ranges.

Table 7.4 Values for the Factor d_2

Sample size, m	d_2	Sample size, m	d_2
2	1.13	14	3.41
3	1.69	15	3.47
4	2.06	16	3.53
5	2.33	17	3.59
6	2.53	18	3.64
7	2.70	19	3.69
8	2.85	20	3.74
9	2.97	21	3.78
10	3.08	22	3.82
11	3.17	23	3.86
12	3.26	24	3.90
13	3.34	25	3.93

Table 7.5 Observations for Samples A to J

Observation times	Sample									
	A	B	C	D	E	F	G	H	I	J
1	10	14	20	10	19	7	10	14	8	13
2	12	17	8	12	20	18	15	12	15	11
3	11	9	11	14	9	15	13	18	10	12
4	15	16	17	8	11	13	9	17	13	17
5	19	20	13	9	9	16	10	11	20	10
6	8	13	20	13	10	15	15	14	17	8
7	7	12	10	7	7	18	16	9	16	15

Table 7.6 Ranges of Samples A–J

	Sample									
	A	B	C	D	E	F	G	H	I	J
Sample range	12	11	12	7	13	11	7	9	12	9

Table 7.7 Mean Values of Samples A–J

	Sample									
	A	B	C	D	E	F	G	H	I	J
Sample mean	11.71	14.43	14.14	10.43	12.14	14.57	12.57	13.57	14.14	12.29

With the aid of data given in Table 7.5, the mean values of samples A–J are tabulated in Table 7.7.

The mean of mean value of Table 7.7 is

$$\bar{\bar{X}} = \frac{(11.71) + (14.43) + (14.14) + (10.43) + (12.14) + (14.57) + (12.57) + (13.57) + (14.14) + (12.29)}{10}$$

$$= 12.999$$

For $m = 7$, the number of observations in each sample, with the aid of Table 7.4 and Eq. (7.33) we get

$$A_2 = \frac{3}{(7)^{1/2}(2.70)} = 0.42$$

By substituting the calculated values for $\bar{\bar{X}}$, A_2, and $\bar{r}$ in Eqs. (7.31)–(7.32), respectively, we get

$$\begin{aligned} \text{UCL}_{\bar{x}} &= \bar{\bar{X}} + A_2\bar{r} \\ &= (12.999) + (0.42)(10.3) \\ &= 17.33 \end{aligned}$$

and

$$\begin{aligned} \text{LCL}_{\bar{x}} &= \bar{\bar{X}} - A_2\bar{r} \\ &= (12.999) - (0.42)(10.3) \\ &= 8.67 \end{aligned}$$

With the aid of calculated values for $\bar{\bar{X}}$, $\text{UCL}_{\bar{x}}$, $\text{LCL}_{\bar{x}}$, and the mean values tabulated in Table 7.7, the $\bar{X}$-chart in Fig. 7.10 was developed. The chart shows that all the mean values of Table 7.7 fall within upper and lower control limits, which means that there is no abnormality.

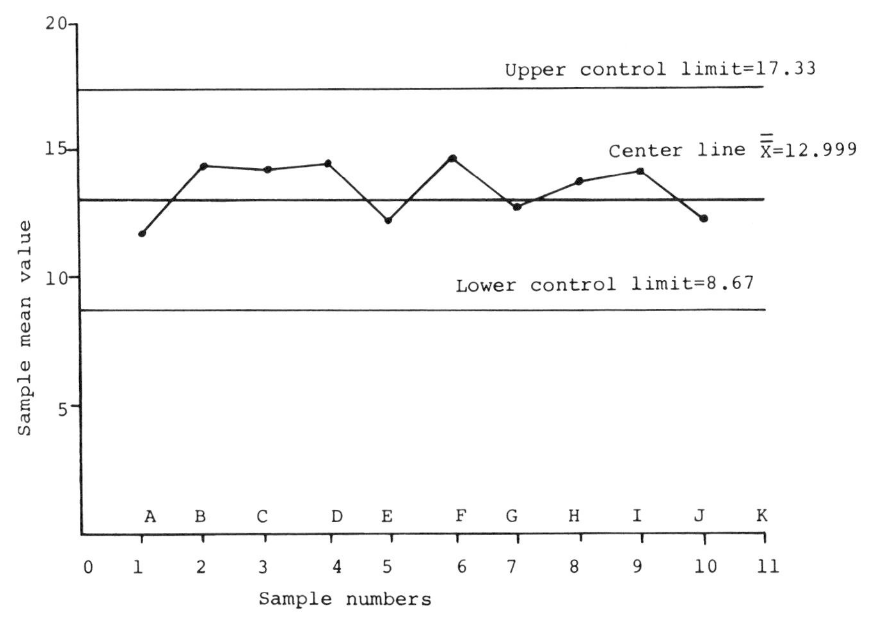

Figure 7.10 The $\bar{X}$-chart.

7.4.3 The R-Charts

The R-charts are known as the control charts for ranges. In this situation, the sample range is used as the variability measure. The range of a sample is equal to the difference between the highest observation value and the lowest observation value, in that sample. The ranges of the samples are plotted on the R-charts.

From Ref. 2, the upper and lower control limits for the R-chart are given by

$$\mathrm{UCL}_r = D_4 \bar{r} \tag{7.34}$$

and

$$\mathrm{LCL}_r = D_3 \bar{r} \tag{7.35}$$

where

$$D_3 \equiv 1 - 3\frac{d_3}{d_2} \tag{7.36}$$

$$D_4 \equiv 1 + 3\frac{d_3}{d_2} \tag{7.37}$$

Table 7.8 Values for the Factor d_3

Sample size, m	d_3	Sample size, m	d_3
2	0.85	14	0.76
3	0.89	15	0.76
4	0.88	16	0.75
5	0.86	17	0.74
6	0.85	18	0.74
7	0.83	19	0.73
8	0.82	20	0.73
9	0.81	21	0.72
10	0.80	22	0.72
11	0.79	23	0.72
12	0.78	24	0.71
13	0.77	25	0.71

and

UCL_r = the upper control limit of the R-chart
LCL_r = the lower control limit of the R-chart
d_3 = the factor whose values are tabulated in Table 7.8
d_2 = the factor whose values are tabulated in Table 7.4

The value of D_3 is taken as zero for the sample with less than seven observations. For various sample sizes, the values of factor d_3 are tabulated [6] in Table 7.8.

Example 7.11 With the aid of specified data in Example 7.10 develop the R-chart.

Table 7.6 exhibits the values of the range for samples A–J. Thus the mean value of ranges shown in Table 7.6 is

$$\bar{r} = 10.3$$

For a sample size equal to seven, with the aid of Tables 7.4 and 7.8 Eqs. (7.36) and (7.37) we get

$$D_3 = 1 - \frac{3d_3}{d_2} = 1 - \frac{(3)(0.83)}{(2.70)} = 0.0778$$

and

$$D_4 = 1 + \frac{3d_3}{d_2} = 1 + \frac{(3)(0.83)}{(2.70)} = 1.9222$$

By utilizing the above calculations in Eqs. (7.34) and (7.35), we get the following values for the upper and lower control limits:

$$LCL_r = (1.9222)(10.3) = 19.799$$

and

$$LCL_r = (0.0778)(10.3) = 0.8013$$

For the above results and calculations of Table 7.6 the R-chart is shown in Fig. 7.11. The chart shows that all the values of Table 7.6 lie within upper and lower control limits, which means that there is no abnormality.

7.4.4 The c-Charts

The c-charts are also known as the control charts for defects per unit. Theoretically these charts are used in situations where the opportunities for defects to occur in an item are large. In other words, these charts are used to control the number of defects in the item. The expressions for the control

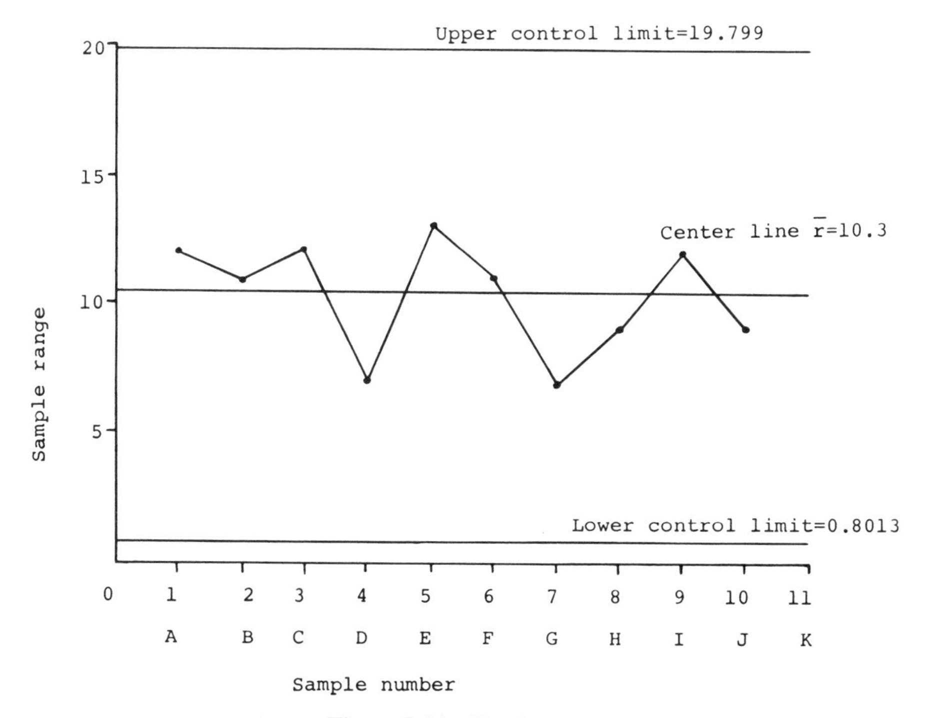

Figure 7.11 The R-chart.

limits are developed with the aid of the Poisson distribution. The mean $\bar{c}$ of the Poisson distribution is given by

$$\bar{c} = \frac{\text{total number of defects}}{\text{total number of items}} \tag{7.38}$$

The standard deviation of the Poisson distribution is

$$\sigma_p = \sqrt{\bar{c}} \tag{7.39}$$

where σ_p is the Poisson distribution standard deviation.

Thus the upper and lower control limits of the c-chart are as follows:

$$UCL_c = \bar{c} + 3\sigma_p \tag{7.40}$$

and

$$LCL_c = \bar{c} - 3\sigma_p \tag{7.41}$$

where UCL_c is the c-chart upper control limit and LCL_c is the c-chart lower control limit.

Example 7.12 Ten electric motors were examined for defects. The number of defects associated with motors 1, 2, 3, 4, 5, 6, 7, 8, 9, and 10 were 5, 4, 10, 7, 10, 8, 8, 5, 6, and 7, respectively. Develop the c-chart.

$$\begin{aligned}\text{Total number of defects} &= 5 + 4 + 10 + 7 + 10 + 8 + 8 + 5 + 6 + 7 \\ &= 70\end{aligned}$$

Total number of electric motors = 10

Thus utilizing the above calculations in Eq. (7.38) results in

$$\bar{c} = \frac{70}{10} = 7 \text{ defects/motor}$$

From Eq. (7.39) we get

$$\sigma_p = \sqrt{7} = 2.6458$$

Using the above calculations in Eqs. (7.40) and (7.41) yields

$$UCL_c = 7 + (3)(2.6458) = 14.9373$$

and

$$LCL_c = 7 - (3)(2.6458) = -0.9374$$

The value of the lower control limit, −0.9374, is impossible. Therefore, it is changed to zero. In other words $LCL_c = 0$. For the above results and the specified data, the c-chart is shown in Fig. 7.12. Figure 7.12 shows that all the data values specified in Example 7.12 fall within the upper and lower control limits, which means that there is no abnormality.

7.5 SUMMARY

This chapter briefly presents the various important aspects of the statistical quality control. It begins by describing briefly the historical aspect of the statistical quality control and then reviews basic statistics. For example, frequency distributions, histograms, range, arithmetic mean, mean deviation, standard deviation, and variance are explained with the aid of numerical examples.

The topic of acceptance sampling is discussed in detail. Advantages and disadvantages of sampling are presented. The lot formation and sampling error are described. Three types of sampling plans are discussed. These are single, double, and multiple sampling plans. The operating characteristic curve and its construction and properties are discussed.

Finally the chapter describes four types of commonly used quality control charts with the aid of solved numerical examples. These charts are

1. The p-chart
2. The $\bar{X}$-chart
3. The R-chart
4. The c-chart

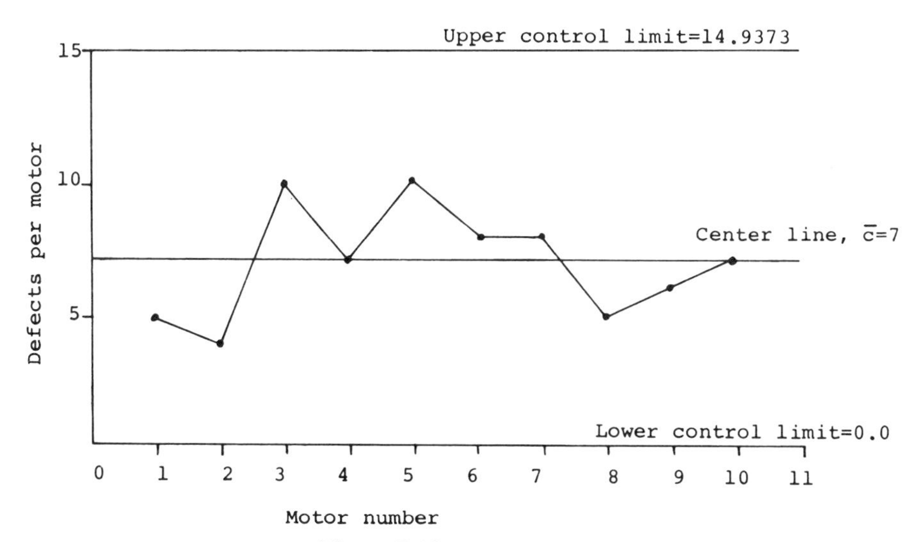

Figure 7.12 The c-chart.

The chapter contains 12 solved numerical examples.

EXERCISES

1. Discuss the history of statistical quality control.
2. Calculate the range of the following data:

   [200, 10, 5, 150, 30, 80, 90, 2, 45]

3. Five identical components were put on the test at time $t=0$. All the components failed during the test. Their failure times were 400, 380, 410, 420, and 370 hr. Compute the value of the mean deviation and the standard deviation.
4. Discuss the advantages and disadvantages of sampling.
5. Explain with the aid of diagrams the following two terms:
   a. Single sampling
   b. Multiple sampling
6. Describe the procedure to construct an operating characteristic curve. Demonstrate it by solving one hypothetical numerical example.
7. Describe in detail the meanings of the following abbreviations:
   a. AQL
   b. LTPD
8. What is the difference between the control charts for attributes and control charts for variables?
9. What are advantages and disadvantages of the control charts?
10. Twenty electric transformers were examined for defects. After the investigation it was established that each transformer contains a certain

Table 7.9 Defects Associated with Transformers

Transformer number	Number of defects per transformer	Transformer number	Number of defects per transformer
1	10	11	12
2	15	12	6
3	8	13	8
4	7	14	10
5	15	15	20
6	6	16	6
7	7	17	10
8	10	18	12
9	11	19	14
10	13	20	10

number of defects. Data for defects associated with each transformer are given in Table 7.9. Develop the c-chart.

REFERENCES

1. J. D. Braverman, *Fundamentals of Statistical Quality Control*, Reston Publishing Co., Reston, Virginia, 1981, p. 6.
2. J. M. Juran, and F. M. Gryna, *Quality Planning and Analysis*, McGraw-Hill, New York, 1980, pp. 407–408.
3. D. H. Besterfield, *Quality Control*, Prentice-Hall, Englewood Cliffs, New Jersey, 1979.
4. G. E. Hayes, and H. G. Romig, *Modern Quality Control*, Bruce: A division of Benziger Bruce & Glencoe, Inc., Encino, California, 1977, pp. 640–641.
5. A. V. Feigenbaum, *Total Quality Control*, McGraw-Hill, New York, 1983, pp. 396.
6. E. S. Buffa, *Operations Management: Problems and Models*, John Wiley & Sons, New York, 1972, pp. 613–631.

8

Applied Quality Control

8.1 INTRODUCTION

The objective of this chapter is to describe briefly various application areas of quality control—in other words, the application of quality control concepts to areas such as computer software, manufacture of electronic products, banking operations, and so on. In the broad sense, the basic quality control methods and procedures remain the same in all the application areas, but the approach and amount of quality control theory application may differ significantly from one industry to another. Furthermore, in some areas one may be concerned with the quality of the service rather than the quality of a product. Today quality control concepts find application in many diverse areas, such as the automotive industry, chemical process industry, metals industry, textile industry, food industry, transportation industry, banking industry, electronic industry, and computer industry. Some specific areas of quality control application are described in the subsequent sections of this chapter.

8.2 QUALITY ASSURANCE OF SOFTWARE

According to Ref. 1, annually $20 billion is spent to develop computer software. Furthermore, in today's environment, it is not unusual to have a computer program with one million instructions. As the computer program size increases so does the problem of software quality assurance. In software work the objective of quality assurance is to assure the quality of software items.

According to Mendis in Ref. 2, many of the software quality assurance programs are linked to the weak comprehension of software quality assurance procedures and techniques.

Some of the reasons are as follows:

1. Software quality assurance needs are treated as secondary.
2. Software quality assurance requirements are defined unclearly.
3. A well-developed methodology is lacking.
4. Personnel are used whose primary interest is not quality. In other words, such personnel treat quality as their secondary interest.

Fischer in Ref. 3 has addressed the software development problem by outlining its symptoms and causes. Some of the symptoms of the software development problem are given in Fig. 8.1.

Similarly, from Ref. 3 some of the causes of the software development problem are as follows:

1. Ineffective management planning, organization, and control
2. Shallow knowledge of user requirements
3. Inadequate programming methods
4. Inadequate design and documentation methods
5. Inadequately defined requirements
6. Ineffective test planning
7. Ineffective communication among concerned groups
8. Inadequate requirements and design verification
9. Ineffective test techniques

8.2.1 Phases of the Software Development Process

According to Ref. 1, the software development process can be broken down into seven phases. These are as follows:

Phase 1. This is known as the software requirement phase. It is probably the most difficult phase, broadly speaking, in which analysis of system requirements are performed and necessary software functions are identified.

Phase 2. This is known as the preliminary design phase. The programming teams are responsible for the preliminary design. In this phase, basically, the software is divided into computer program elements and necessary performance needs assigned to them. Furthermore, memory and timing are established.

Phase 3. This is known as the detail design phase; where, basically, computer program elements are subdivided further into routines to which performance needs are assigned.

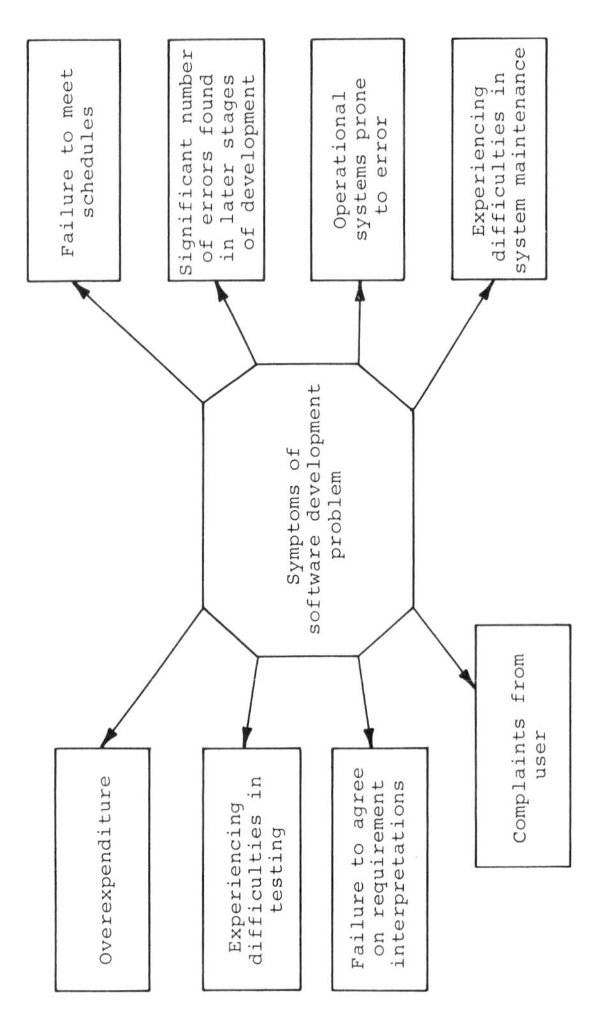

Figure 8.1 Symptoms of the software development problem.

Phase 4. This is known as the code, debug, and unit test phase. Actually, this is the phase in which software is produced and associated actions are performed.

Phase 5. This is known as the development testing phase, where earlier tested routines are integrated into computer program elements. Furthermore, this phase is concerned with the verification of outlined functional capabilities of computer program elements.

Phase 6. This is known as the integration testing phase. This phase is concerned with the verification of computer program element level requirements, for example, proper data routing, element interfaces, and proper sequencing of elements. Usually, this phase is conducted by an independent group. The end result of this phase is totally integrated software.

Phase 7. This is the final phase of the software development process and is known as acceptance testing. The phase is concerned with the verification of system level functional requirements. Examples of such requirements are ability to handle successfully the overall input load and overall timing. Also in this phase, usually a demonstration is conducted for the customer to show software validity.

8.2.2 Functions of Software Quality Assurance

These functions may vary from one project to another. However, according to Ref. 4 there are basically eight functions, as shown in Fig. 8.2.

Functions of software quality assurance given in Fig. 8.2 are considered to be self-explanatory; therefore, they are not described here. However, if a need arises for detailed explanation, the reader should consult Ref. 4.

8.2.3 Advantages of Software Quality Assurance

The many advantages of software quality assurance program were realized at TRW, Inc. [4]. Most of them are as follows:

1. There is less project risk because of systematic testing and better requirements traceability.
2. There is quality assurance record centralization.
3. Meeting of contractural requirements of deliverable units is assured by an independent team.
4. Through audits and reviews the quality assurance program enhanced management visibility in the development process.
5. It is helpful in enforcing software standards.

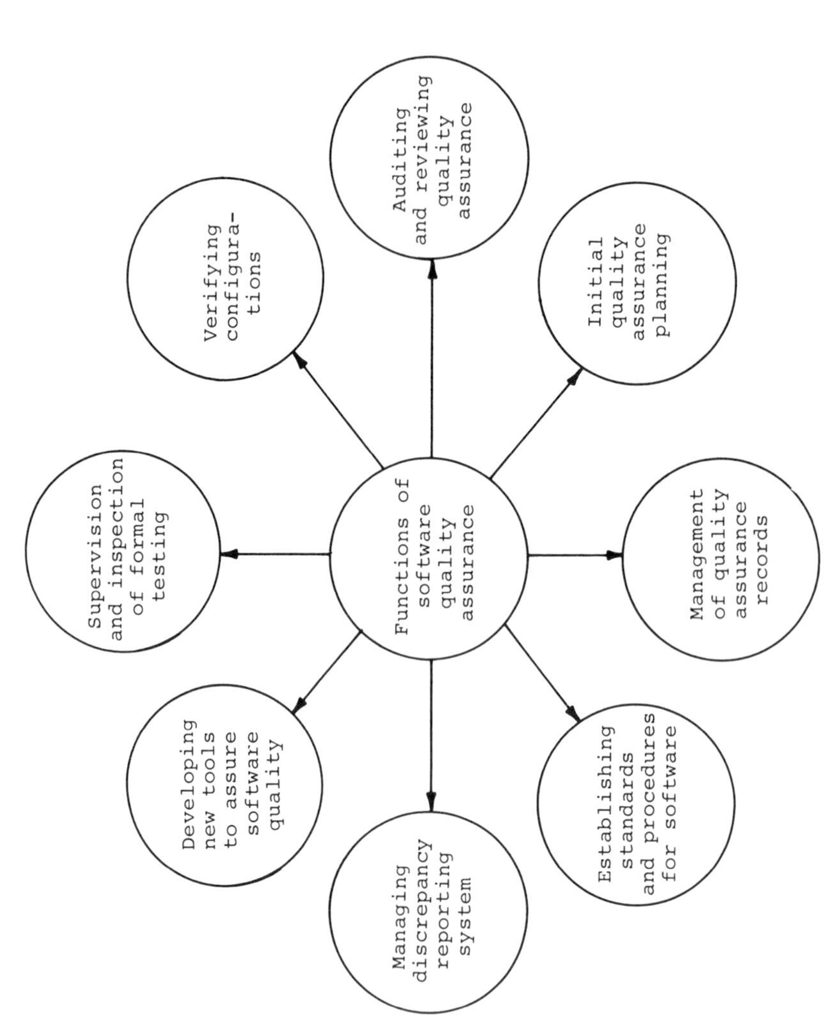

Figure 8.2 Functions of software quality assurance.

8.2.4 Quality Assurance Guidelines

This section briefly outlines guidelines extracted from lessons learned at TRW, Inc. [4], with respect to the implementation of a quality assurance program. In addition, some general guidelines are included. Some of them are as follows:

1. A strong quality assurance activity is one factor in developing reliable software.
2. The advantages accrued from the software quality assurance program are worth their cost.
3. Announce well before auditing the intention to audit the quality assurance software program.
4. The auditing should be directed to ascertaining whether the activity in question is performed correctly rather than to finding the problems.
5. During the proposal and contract definition phase, there must be satisfactory participation from quality assurance.
6. Make sure that the first audit is conducted in the early stage of the development process. This way enough time will be left for corrective measures.
7. Develop a checklist for audits.
8. It is simpler to train personnel with software background in quality assurance than the other way around (i.e., to train personnel with the knowledge of quality assurance only).

8.3 A SYSTEM TO ASSURE QUALITY OF INTEGRATED CIRCUITS

This section briefly summarizes a procedure from Ref. 5 to assure quality of the integrated circuits. The block diagram of phases involved to develop and market the product is shown in Fig. 8.3. Thus, to produce integrated circuits of satisfactory quality, the quality assurance activities have to be performed during all phases outlined in Fig. 8.3.

From Ref. 5 the various quality assurance activities performed during various phases shown in Fig. 8.3 are discussed in following sections:

(a) Quality control at the development level: Here, one is concerned with areas such as establishing a system so that all engineers can participate effectively in quality assurance, taking the market survey, and collecting necessary data.

(b) Quality control of components and materials: These are those items which are going to be used in the end product. Thus, special attention

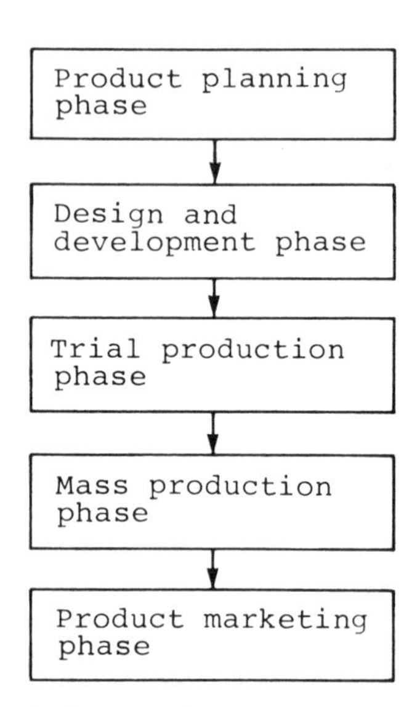

Figure 8.3 Block diagram of the product development phases.

must be given to the quality of the procured items. The quality of procured items can be assured through an effective vendor control system, vendor auditing, receiving inspection, and so on.

(c) Quality control of products in production lines: At this stage of the product, special attention is to be given in the initial stage of the production with respect to failure modes of integrated circuits. Furthermore, to assure the quality of the product at this stage, activities such as use of quality circles with respect to workmanship, setting up quality standards, taking necessary corrective actions to correct abnormalities, making use of necessary statistical techniques, controlling the production line equipment, and taking an interest in the process history are to be performed effectively.

(d) Quality control of lots: To assure the quality of integrated circuit lots, various lot quality assurance tests are to be conducted—for example, steady-state life test, high-temperature and high-humidity life test, high-temperature storage life test, and reliability test. Furthermore, physical appearance, construction, and electrical characteristic inspections are to be performed. Stock and shipment inspection is also a part of the lot quality assurance test.

(e) Reliability assurance of integrated circuits: The reliability tests are conducted in addition to the lot quality assurance test. Various reliability tests are performed to confirm reliability at the specified frequencies. Results

of these tests also become useful in establishing reliability control plans and procedures, developing failure analysis methods, studying failure mechanism, reliability data collection, etc.

(f) Quality assurance in service activities: This is an important component of the overall quality assurance of the integrated circuits. Examples of service activities are handling complaints and grasping the quality of the product in the field. Performing activities such as these effectively requires properly planned procedures. Thus, proper care must be given when planning such procedures. Service activities are also concerned with analyzing the field data and initiating correct measures as necessary.

8.4 QUALITY CONTROL IN BANKING SYSTEMS

Banks are one of the late comers to take advantage of the quality control discipline. Nowadays some banks are beginning to look into quality control to improve their productivity. Banking operations have same characteristics like other service industries. Examples of such characteristics are labor intensiveness, small processing time, and fast turn-around time. Similarly, just as in any other service industry, the bank customers are most concerned with "service" performance, reliability of service, availability of service, and so on. According to Ref. 6, in banks computers are the prime source of quality-associated problems. These problems may be classified into four categories, namely, operating personnel, computer software, computer hardware, and control systems. Low productivity results from such problems—for example, in terms of poor work efficiency, costs of delays, and overtime.

8.4.1 Implementation of Quality Control Program for Bank Operations

This section summarizes steps to implement a quality control program for bank operations. These steps are explained in detail for service industries in Ref. 7. According to Langevin [8], the same steps can also be applied to bank operations. His steps are listed below:

1. Perform analysis of product needs.
2. Conduct analysis of the production process.
3. Establish acceptance needs such as service criteria, quality, timeliness.
4. Develop necessary checkpoints for inspection.
5. Report necessary service data and quality.
6. Examine reports.
7. Review performance.

8. Perform analysis of errors and trends.
9. Analyze costs.
10. Develop performance improvement strategies.
11. Establish policies and plans for corrective measures.
12. Keep track of implementation.
13. Perform audits of procedural compliance and keep track of integrity of data.

8.4.2 Useful Techniques for Bank Operations

The well-established techniques for product industries can also be applied to bank operations. Some of these are described below:

1. *Pareto analysis*: In bank operations Pareto analysis is also useful to find out what is the most pressing problem. For example, which one of the following four cause the most problem: (a) computer operating personnel; (b) computer software; (c) control; (d) hardware of the computer?
2. *Control charts*: These are also useful to monitor trends and performance levels over a time period in the bank operations.
3. *Cause-and-effect diagrams*: This is another useful technique having applications in bank operations, because many generic problems evolve from personnel, measurement process, machines, materials, or techniques.
4. *Failure mode and effects analysis*: This technique is widely used to perform reliability analysis of engineering systems. In bank operations it is useful to identify possible failure modes of computer systems.
5. *Design of experiment*: In product industries, this is used to screen a small number of interactions and factors out of several factors which are causing problems. Therefore, in bank operations it can also be applied to sort out interactions of computer hardware, computer software, operating personnel, and control procedures.
6. *Quality information system*: In the product industries this is used to collect quality-associated data concerned with warranties, scrap, and so on. In bank operations this type of system can also play an important role. For example, the quality information system can be used to keep track of payments or overtime, delays, quality-related indices, etc.

8.5 QUALITY CONTROL IN THE TEXTILE INDUSTRY

Increasing consumer demand, costs, and competition have led the way for the application of quality control concepts in the textile industry. Therefore, this section briefly outlines a quality control program to produce better-quality textile products. According to Hart in Ref. 9, the four-component quality control program shown in Fig. 8.4 can be used throughout the textile industry.

The components of the quality control program for textile industry shown in Fig. 8.4 are described in detail below:

1. *Control of incoming materials*: This can be accomplished by auditing suppliers, classifying suppliers, incoming inspection, developing testing and rejection program, and so on.
2. *Control of outgoing products*: This is accomplished through the followings:
   a. Audit the returns from customers and notify the appropriate people who are responsible for the failures.
   b. Audit the quality of the outgoing packs and take necessary corrective measures.
   c. Determine "critical few" packing units and packers and inspectors through auditing.
   d. Follow up with management to see if appropriate corrective measures have been carried out.
3. *In-process controls*: These are accomplished through the following steps:
   a. Identify "significant few" manufacturing areas that cause most quality-related problems.
   b. Identify "significant few" defects that cause most of the off-quality.
   c. Set up effective in-process inspection whenever it is necessary.
   d. To reduce off-quality, do laboratory testing of materials as necessary.
   e. Provide assistance to manufacturing with special projects

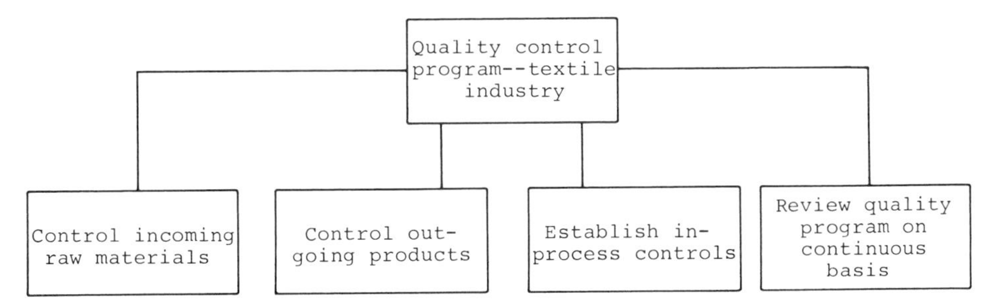

Figure 8.4 Quality control program for textile industry.

concerned with finding the best techniques and procedures of operation to reduce scrap and off-quality.

f. Follow up with management to see if appropriate corrective measures have been carried out.

8.6 SUMMARY

This chapter briefly describes four important areas of applied quality control. These are quality assurance of software, quality control of integrated circuits, quality control in banking system, and quality control in textile industry.

The first topic discussed in the chapter is concerned with software quality assurance. Symptoms and causes of the software development problem are briefly discussed. Seven phases of the software development problem are described. Eight functions of software quality assurance are briefly explained. Benefits of the software quality assurance and quality assurance guidelines are listed.

The next topic discussed in the chapter is concerned with quality assurance of integrated circuits. Various steps to assure quality of integrated circuits are summarized.

Quality control in banks is another topic which is also briefly described in the chapter. Various steps are listed to implement a quality control program for bank operations. Useful quality control techniques for bank operations are briefly discussed.

The last topic discussed in the chapter is quality control in the textile industry. A universal textile quality control program is presented.

EXERCISES

1. Discuss the need for having a software quality assurance program.
2. What are the functions of software quality assurance.
3. Describe phases of the software development process.
4. What is the main difference between software quality assurance and software reliability?
5. What are the benefits and drawbacks of having a software quality assurance program?
6. Discuss briefly the following:
   a. Pareto analysis
   b. Cause-and-effect diagrams
   c. Failure modes and effects analysis
   d. Design of experiment

REFERENCES

1. J. McKissick, Quality Control of Computer Software, *Proceedings of the American Society for Quality Control Annual Conference* (*Transactions*), 1977, pp. 391–398.
2. K. S. Mendis, A Software Quality Assurance Program for the 80's, *Proceedings of the American Society for Quality Control Annual Converence (Transactions)*, 1980, pp. 379–388.
3. K. F. Fischer, A Methodology for Developing Quality Software, *Proceedings of the American Society for Quality Control Annual Conference (Transactions)*, 1979, pp. 364–371.
4. K. F. Fischer, A Program for Software Quality Assurance, *Proceedings of the American Society for Quality Control Annual Conference (Transactions)*, 1978, pp. 333–340.
5. M. Takanashi, Quality Assurance System for the Integrated Circuit, *Quality Assurance: Methods, Management and Motivation*, edited by H. J. Bajaria, Society of Manufacturing Engineers, Dearborn, Michigan, 1981, pp. 191–196.
6. K. E. C. Anyanonu, and H. J. Bajaria, Quality Control in Banks—Why, Where and How, *Proceedings of the American Society for Quality Control Annual Conference (Transactions)*, 1980, pp. 307–312.
7. R. G. Langevin, Quality Control in the Service Industries, An American Management Associations (AMA) Management Briefing, American Management Associations, 135 West 50th Street, New York, 1977.
8. R. G. Langevin, Quality Control in Bank Operations, *Proceedings of the American Society for Quality Control Annual Conference (Transactions)*, 1983, pp. 131–135.
9. W. C. Hart, Universal Textile Quality Control—Why, Who, and How, *Proceedings of the American Society for Quality Control Annual Conference (Transactions)*, 1979, pp. 28–34.

9

Introduction to Reliability

9.1 INTRODUCTION

Reliability is a relatively new discipline whose history goes back to World War II. The conception of this field was basically due to the sophistication, automation, and complexity of modern systems. After the end of World War II, several studies were conducted by the United States Armed Forces which exhibited the pressing need to improve reliability of military equipment. For example, the following were some of the findings [1] of these studies:

1. The Air Force study indicated that the cost to maintain equipment in the field was about ten times higher than the acquisition cost.
2. The electronic equipment was nonoperative about 70 percent of the time during maneuvers, according to the study conducted by the Navy.

Since the end of the war many advances have been made in the reliability discipline. Today, when complex engineering systems are designed, reliability is usually considered as one of the design parameters. However, factors such as the following impose constraints on efforts to improve reliability [2]:

1. Finished product's volume
2. Uncertainty of product's field use environments
3. Design effort cost
4. Components' manufacturing cost
5. Final product's weight

6. Logistics
7. Availability of necessary personnel

This chapter describes various introductory aspects of reliability in the subsequent sections.

9.2 RELIABILITY BASICS

This section briefly discusses cumulative distribution function, probability density function, reliability function, hazard rate, mean time to failure, and variance.

9.2.1 Cumulative Distribution Function

The cumulative distribution function, $F_c(t)$, of a continuous random variable is given by

$$F_c(t) = \int_{-\infty}^{t} f(x)\, dx \tag{9.1}$$

where $f(t)$ is the probability density function and t is time.

Alternative equations to obtain the cumulative distribution function are as follows:

$$F_c(t) = 1 - R(t) \tag{9.2}$$

and

$$F_c(t) = 1 - \exp\left[-\int_0^t h(x)\, dx\right] \tag{9.3}$$

where $R(t)$ is the reliability function and $h(t)$ is the hazard rate.

Example 9.1 Failure times of an electronic component are described by the following probability density function:

$$f(t) = \lambda e^{-\lambda t} \tag{9.4}$$

where λ is the component constant failure rate and is equal to 0.002 failures/hr. Calculate the failure probability of the electronic component for a 20-hr mission. Assume that the component starts operating at time $t = 0$.

Substituting Eq. (9.4) into Eq. (9.1) yields

$$\begin{aligned} F_c(t) &= \int_0^t (0.002)e^{-(0.002)x}\, dx \\ &= 1 - e^{-(0.002)t} \end{aligned} \tag{9.5}$$

For $t = 20$ hr, from Eq. (9.5) we get

$$F_c(20) = 1 - e^{-(0.002)(20)} = 0.0392$$

For the specified mission time, the failure probability of the component is 0.0392.

9.2.2 Probability Density Function

This is defined as

$$f(t) = \frac{dF_c(t)}{dt} \tag{9.6}$$

In addition, the following relationships can also be used to obtain the probability density function:

$$f(t) = h(t)R(t) \tag{9.7}$$

and

$$f(t) = -\frac{dR(t)}{dt} \tag{9.8}$$

Example 9.2 A mechanical component's hazard rate and reliability function, respectively, are defined by

$$h(t) = \frac{\theta t^{\theta-1}}{\mu^{\theta}} \tag{9.9}$$

and

$$R(t) = \exp\left[-\left(\frac{t}{\mu}\right)\right]^{\theta} \tag{9.10}$$

where μ is the scale parameter, θ is the shape parameter, and t is time. Obtain the failure probability density function of the component.

By substituting Eqs. (9.9)–(9.10) into Eq. (9.7) leads to

$$f(t) = h(t)R(t) = \frac{\theta t^{\theta-1}}{\mu^{\theta}} \exp\left[-\left(\frac{t}{\mu}\right)^{\theta}\right]$$

$$\text{for } \theta > 0, \quad \mu > 0, \quad t \geq 0 \tag{9.11}$$

9.2.3 Reliability Function

This is defined as

$$R(t) = \int_t^{\infty} f(x)\, dx \qquad (9.12)$$

where $R(t)$ is the reliability function of a component, a subsystem or a system.

Other equations also used to obtain the reliability function, $R(t)$, are as follows:

$$R(t) = 1 - F_c(t) = 1 - \int_0^t f(x)\, dx \qquad (9.13)$$

and

$$R(t) = \exp\left[-\int_0^t h(x)\, dx\right] \qquad (9.14)$$

Example 9.3 An electric motor's constant failure rate, λ, is 0.0004 failures/hr. Calculate the motor reliability for a 150-hr mission. Assume that the motor starts operating at time $t = 0$. Then

$$h(t) = \lambda = 0.004 \text{ failures/hr}$$

Substituting the above data into Eq. (9.14) results in

$$R(t) = \exp\left[-0.0004 \int_0^t dx\right] = \exp[-(0.0004)t]$$

Thus from the above equation, the reliability of the motor for a 150-hr mission is

$$R(150) = \exp[-(0.004)(150)]$$
$$= 0.9418$$

This means the reliability of the motor is 0.9418 for the specified period of operation.

9.2.4 Hazard Rate

The hazard rate (or time-dependent failure rate) $h(t)$ of a component, a subsystem, or a system is given by

$$h(t) = \frac{f(t)}{R(t)} \qquad (9.15)$$

The hazard rate function of an item can also be obtained from the following relationships:

$$h(t) = -\frac{1}{R(t)} \frac{dR(t)}{dt} \tag{9.16}$$

and

$$h(t) = \frac{f(t)}{1 - F_c(t)} \tag{9.17}$$

Example 9.4 The reliability function $R(t)$ of an aircraft engine is given by

$$R(t) = \exp\left[-\left(\frac{t}{\mu}\right)^2\right] \tag{9.18}$$

where μ is the scale parameter and t is time. Obtain an expression for the aircraft's hazard rate.

Differentiating Eq. (9.18) with respect to time t results in

$$\frac{dR(t)}{dt} = -\frac{2t}{\mu^2} \exp\left[-\left(\frac{t}{\mu}\right)^2\right] \tag{9.19}$$

Substituting Eqs. (9.18) and (9.19) into Eq. (9.16) yields

$$\begin{aligned} h(t) &= -\frac{1}{\exp[-(t/\mu)^2]} \left\{-\frac{2t}{\mu} \exp\left[-\left(\frac{t}{\mu}\right)^2\right]\right\} \\ &= \frac{2t}{\mu^2} \end{aligned} \tag{9.20}$$

9.2.5 Mean Time to Failure

The mean time to failure (MTTF) of an item is defined as

$$\text{MTTF} = \int_0^\infty R(t)\, dt \tag{9.21}$$

An alternative approach to obtain the same result is to find the expected value of the failure probability density function. Thus, the expected value $E(t)$ of a continuous random variable is given by

$$E(t) = \text{MTTF} = \int_0^\infty t f(t)\, dt \tag{9.22}$$

Example 9.5 The reliability function of an electric fan is described by

$$R(t) = e^{-\lambda t} \tag{9.23}$$

The constant failure rate of the fan is 0.0008 failures/hour. With the aid of Eq. (9.23) calculate the value of the electric fan's mean time to failure.

Utilizing Eq. (9.23) in Eq. (9.21) leads to

$$\text{MTTF} = \int_0^\infty e^{-\lambda t}\, dt = \left(-\frac{e^{-\lambda t}}{\lambda}\right)_0^\infty$$

$$= \frac{1}{\lambda} \tag{9.24}$$

Substituting the specified failure rate data into Eq. (9.24) yields

$$\text{MTTF} = \frac{1}{0.0008} = 1250 \text{ hr}$$

Thus, the mean time to failure of the electric fan is 1250 hr.

9.2.6 Variance

This is used to measure the dispersion of a distribution and is defined as

$$\sigma^2 = \int_0^\infty (t - \text{MTTF})^2 f(t)\, dt \tag{9.25}$$

where σ^2 denotes variance.

By expanding Eq. (9.25) we get

$$\sigma^2 = \int_0^\infty t^2 f(t)\, dt - 2(\text{MTTF}) \int_0^\infty t f(t)\, dt + (\text{MTTF})^2 \int_0^\infty f(t)\, dt \tag{9.26}$$

From Eq. (9.22) we have

$$\text{MTTF} = \int_0^\infty t f(t)\, dt \tag{9.27}$$

Furthermore, by definition the total area under the plot of a probability density function is always equal to unity, that is,

$$\int_0^\infty f(t)\, dt = 1 \tag{9.28}$$

With the aid of relationships (9.27)–(9.28), Eq. (9.26) simplifies to

$$\sigma^2 = \int_0^\infty t^2 f(t)\, dt - 2(\text{MTTF})^2 + (\text{MTTF})^2$$

$$= \int_0^\infty t^2 f(t)\, dt - (\text{MTTF})^2 \tag{9.29}$$

The above equation is used to compute variance.

Example 9.6 The failure probability density function of the Weibull distribution from Eq. (9.11) is

$$f(t) = \frac{\theta t^{\theta-1}}{\mu^\theta} \exp\left[-\left(\frac{t}{\mu}\right)^\theta\right] \tag{9.30}$$

Obtain an expression for the variance.

Substituting Eq. (9.30) into Eq. (9.27) and integrating results in

$$\text{MTTF} = \frac{\theta}{\mu^\theta} \int_0^\infty t^\theta \exp\left[-\left(\frac{t}{\mu}\right)^\theta\right] dt$$

$$= \mu\Gamma\left(\frac{1}{\theta} + 1\right) \tag{9.31}$$

where

$$\Gamma(\beta) = \int_0^\infty t^{\beta-1} e^{-t}\, dt \tag{9.32}$$

Equation (9.32) is the definition of the gamma function. Substituting Eqs. (9.30) and (9.31) into Eq. (9.29) yields

$$\theta^2 = \frac{\theta}{\mu^\theta} \int_0^\infty t^{\theta+1} \exp\left[-\left(\frac{t}{\mu}\right)^\theta\right] dt - \left[\mu\Gamma\left(\frac{1}{\theta} + 1\right)\right]^2$$

$$= \mu^2\Gamma\left(\frac{\theta+2}{\theta}\right) - \left[\mu\Gamma\left(\frac{1}{\theta} + 1\right)\right]^2 \tag{9.33}$$

9.3 PROBABILITY DISTRIBUTION FUNCTIONS USED IN RELIABILITY PREDICTION

This section briefly discusses selective probability distributions used in the reliability analysis of engineering systems. Therefore, the binomial, Poisson, bathtub, exponential, and Weibull distributions are described below.

9.3.1 The Binomial Distribution

This distribution belongs to the family of discrete distributions. The distribution is used in a situation where a trial must result in either success or failure. The density function is defined as

$$f(k) = \binom{m}{k} p^k q^{m-k} \quad \text{for} \quad 0 \le p \le 1, \quad q = 1 - p, \quad k = 1, 2, 3, \ldots, m \tag{9.34}$$

$$\binom{m}{k} = \frac{m!}{k!(m-k)!} \tag{9.35}$$

where

p = the probability of success
q = the probability of failure
m = the number of independent trials and $m! = m(m-1)(m-2)(m-3)\cdots 1$
k = the successes in m trials
$f(k)$ = the probability of exactly k successes

The binomial cumulative distribution function is given by

$$F(k) = \sum_{j=0}^{k} \binom{m}{j} p^j q^{(m-j)} \tag{9.36}$$

where $F(k)$ is the probability of k or fewer successes in m trials.

In reliability analysis the following expression is frequently used to represent the binomial distribution:

$$(p + q)^m \tag{9.37}$$

Expressions for mean and variance [3] associated with the binomial distribution are as follows, respectively:

$$\mu = mp \tag{9.38}$$

and

$$\sigma^2 = pqm \tag{9.39}$$

where μ is the mean.

Example 9.7 A coin is tossed four times. Calculate the probability of having exactly three tails. In each toss the probability of having either head or tail is 0.5.

In this example the data for the elements of the binomial probability density function are as follows:

$$m = 4, \qquad k = 3, \qquad p = q = 0.5$$

Substituting the above data into Eqs. (9.34)–(9.35) results in

$$f(3) = \left[\frac{4!}{3!(4-3)!}\right](0.5)^3(0.5)^{4-3}$$
$$= 0.25$$

Thus the probability of having exactly three tails is 0.25.

Example 9.8 An aircraft is composed of three active and identical engines in parallel. All engines fail independently. At least one engine must function normally for the aircraft to fly successfully. The probability of success of an engine is 0.8. Calculate the probability of the aircraft crashing. Assume that an engine can only be in two states, i.e., operating normally or failed.

Since $p + q = 1$, the probability of failure of an engine is given by

$$q = 1 - p = 1 - 0.8 = 0.2$$

In this example, the data for the elements of equation (9.36) are as follows:

$$m = 3, \qquad k = 0, \qquad p = 0.8, \qquad q = 0.2$$

By substituting the above data into Eq. (9.36), we get

$$F(0) = \binom{3}{0}(0.8)^0(0.2)^3$$
$$= 0.008$$

Thus, the probability of the aircraft crashing is 0.008.

Example 9.9 An engineering company manufactures a specific type of electric switch. A sample of 50 switches was taken. The probability of a defective switch is 0.05. Calculate the expected value and variance for the distribution of defective switches in the specified sample.

Substituting the specified data into Eqs. (9.38) and (9.39), respectively, leads to

$$\mu = (50)(0.05) = 2.5$$

(in the sample of 50 switches we can expect 2.5 of them to be defective) and

$$\sigma^2 = (0.05)(1 - 0.05)(50)$$
$$= 2.38$$

9.3.2 The Poisson Distribution

This is another distribution which belongs to the family of discrete distributions. The failure density function of the Poisson distribution is given by

$$f(n) = \frac{(m)^n}{n!} e^{-m}, \qquad n = 0, 1, 2, 3; \quad m > 0 \tag{9.40}$$

$$m = \lambda t$$

where m is the mean, $f(n)$ is the probability of n number of failures during time t, and λ is the mean failure rate.

The cumulative distribution function is given by

$$F_c(n) = \sum_{i=0}^{n} \frac{m^i e^{-m}}{i!}$$

$$= \sum_{i=0}^{n} \frac{(\lambda t)^i e^{-\lambda t}}{i!} \tag{9.41}$$

The expected value of the distribution is

$$m = \lambda t \tag{9.42}$$

Finally, the variance is given by

$$\sigma^2 = \lambda t \tag{9.43}$$

Example 9.10 A sample contains 1500 units of an engineering product. The failure probability of a unit is 0.0005. Calculate the probability of 5 units failing out of the entire sample.

In this example the failure probability of the unit is very small and the sample is large. Thus, the Poisson probability function will give almost the same result as if we had used the binomial probability function.

Thus, substituting the specified data into Eq. (9.40) leads to

$$f(5) = \frac{(m)^5}{5!} e^{-m}$$

$$= \frac{[(0.0005)(1500)]^5}{5!} e^{-[(0.0005)(1500)]}$$

$$= 0.00093$$

From the above result we can conclude that the probability of 5 units failing is 0.00093.

9.3.3 Bathtub Distribution

The probability density function $f(t)$, the cumulative distribution function $F_c(t)$, and hazard rate $h(t)$ are defined, respectively, as follows:

$$f(t) = \theta\mu(\mu t)^{\theta-1} \exp[-(e^{(\mu t)^\theta} - \mu t^\theta - 1)] \quad \text{for} \quad \theta > 0, \quad \mu > 0, \quad t \geq 0 \tag{9.44}$$

$$F_c(t) = 1 - \exp[-(e^{(\mu t)^\theta} - 1)] \tag{9.45}$$

and

$$h(t) = \theta\mu(\mu t)^{\theta-1} e^{(\mu t)^\theta} \tag{9.46}$$

where t is time, μ is the scale parameter, and θ is the shape parameter.

From Eq. (9.2), the reliability function $R(t)$ is given by

$$R(t) = 1 - F_c(t) \tag{9.47}$$

Substituting Eq. (9.45) into Eq. (9.47) yields

$$R(t) = \exp[-(e^{(\mu t)^\theta} - 1)] \tag{9.48}$$

At $\theta = 0.5$ the plots of Eq. (9.46) follow a similar shape to the one shown in Fig. 9.1. The hazard rate curve shown in this figure is known as the "bathtub hazard rate curve" because its shape resembles a typical bathtub. When predicting an item's reliability, it is frequently assumed that the hazard rate of such an item follows the bathtub curve. It is very much true in the case of electronic components. As shown in Fig. 9.1 the curve has three distinct parts. Part I shows the decreasing hazard rate. This region is known as the

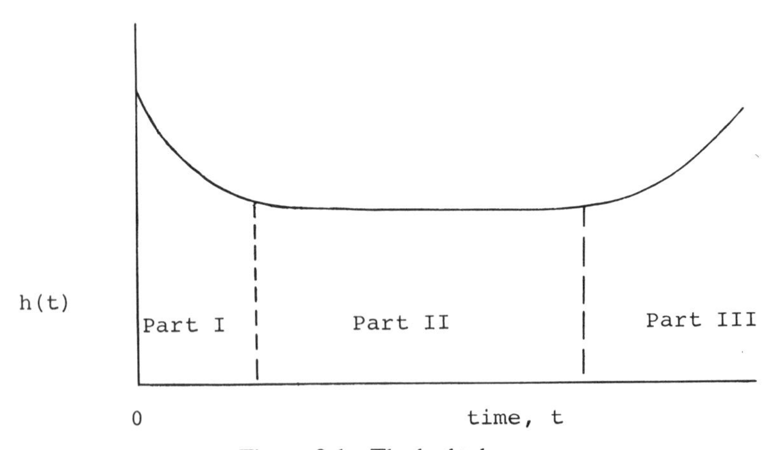

Figure 9.1 The bathtub curve.

infant mortality period. However, various other names are also given to this region. In this period, the failures occur due to design or manufacturing defects. Part II of the curve is known as the constant failure rate region or the useful life period of the item. During this period the failures occur unpredictably. Finally, Part III of the curve is known as the "wear-out period." Failures during this period increase because the item has bypassed its useful operating life.

Frequently in reliability evaluation of an item only the middle portion of the bathtub curve is taken into consideration. In other words it is assumed that the item's failure rate is constant. This means the failure times of the item are exponentially distributed.

9.3.4 Exponential Distribution

This is the most widely used statistical distribution in reliability engineering. The probability density function, cumulative distribution function, and hazard rate, respectively, are defined as follows:

$$f(t) = \lambda e^{-\lambda t}, \qquad t \geq 0 \tag{9.49}$$

where λ is the parameter (i.e., constant failure rate of an item) and t is time.

$$F_c(t) = 1 - e^{-\lambda t} \tag{9.50}$$

and

$$h(t) = \lambda \tag{9.51}$$

By substituting Eq. (9.50) into Eq. (9.47), the resulting reliability function associated with the exponential distribution is

$$R(t) = e^{-\lambda t} \tag{9.52}$$

From Eq. (9.24), the mean time to failure is

$$\text{MTTF} = \frac{1}{\lambda} \tag{9.53}$$

The following expression for the variance is obtained by substituting Eqs. (9.49) and (9.53) into Eq. (9.29):

$$\sigma^2 = \lambda \int_0^\infty t^2 e^{-\lambda t}\, dt - \left(\frac{1}{\lambda}\right)^2$$

$$= \frac{2}{\lambda^2} - \frac{1}{\lambda^2} = \frac{1}{\lambda^2} \tag{9.54}$$

Example 9.11 An electronic computer's constant failure rate is 0.0007 failures/hr. Compute the reliability of the computer for a 300-hr operation. Assume that the computer starts operating at time $t = 0$.

Utilizing the specified data in Eq. (9.52) leads to

$$R(300) = e^{-(0.0007)(300)} = 0.8106$$

The reliability of the computer for the specified mission time is 0.8106.

9.3.5 Weibull distribution

This is another distribution which belongs to the family of continuous distributions. The distribution is named after its originator, Professor W. Weibull, who reported its existence in 1951 [4]. The probability density function, cumulative distribution function, and hazard rate, respectively, are

$$f(t) = \frac{\theta}{\mu^\theta} t^{\theta-1} \exp\left[-\left(\frac{t}{\mu}\right)^\theta\right] \quad \text{for} \quad \theta > 0, \quad \mu > 0, \quad t \geq 0 \tag{9.55}$$

$$F_c(t) = 1 - \exp\left[-\left(\frac{t}{\mu}\right)^\theta\right] \tag{9.56}$$

and

$$h(t) = \frac{\theta t^{\theta-1}}{\mu^\theta} \tag{9.57}$$

The parameters θ and μ are defined in Example 9.2. From Eq. (9.10), the reliability function is given by

$$R(t) = \exp\left[-\left(\frac{t}{\mu}\right)^\theta\right] \tag{9.58}$$

Similarly, from Eqs. (9.31) and (9.33), the MTTF and σ^2, respectively, are given by

$$\text{MTTF} = \mu\Gamma\left(\frac{1}{\theta} + 1\right) \tag{9.58}$$

and

$$\theta^2 = \mu^2\Gamma\left(\frac{\theta + 2}{\theta}\right) - \left[\mu\Gamma\left(\frac{1}{\theta} + 1\right)\right]^2 \tag{9.59}$$

Example 9.12 Failure times of a mechanical component are described by a Weibull distribution. The estimated values of μ and θ are 1000 hr and $\theta = 2$, respectively. Calculate the component reliability for a 200-hr operational period. Assume that the component starts operating at time $t = 0$.

Utilizing the specified data in Eq. (9.58) results in

$$R(50) = \exp\left[-\left(\frac{200}{100}\right)^2\right]$$

$$= 0.9608$$

Thus, the mechanical component reliability for the specified mission is 0.9608.

Example 9.13 With the aid of given data in example 9.12, calculate the mean time to failure of the mechanical component.

For the specified data, from Eq. (9.58), the component

$$\text{MTTF} = (1000)\Gamma\left(\frac{1}{2} + 1\right)$$

$$= (1000)\Gamma(1.5)$$

$$= 886.23 \text{ hr}$$

9.4 RELIABILITY MODELS

This section is concerned with the reliability evaluation of well-known networks such as series and parallel. In all these networks, it is assumed that each element's reliability is constant. Nevertheless, the time-dependent analyses are presented in Chapter 11.

9.4.1 Series Model

This model represents a system with m independent elements connected in series. It means that if any one of the elements fails, the entire series system fails.

The reliability expression for the series model is obtained from Eq. (9.37) as follows:

$$(p + q)^m = (R + q)^m = R^m + mR^{m-1}q + \frac{m(m-1)}{2!}R^{m-2}q^2 + \cdots$$

$$+ \frac{m(m-1)\cdots(m-i+1)}{i!}R^{m-i}q^i + \cdots + q^m \quad (9.60)$$

where

R = the reliability of the element and is equal to p

q = the failure probability of the element

m = the number of elements

R^m = the reliability of m elements operating successfully

$mR^{m-1}q$ = the probability of exactly $(m-1)$ elements operating successfully

$\frac{m(m-1)}{2!}R^{m-2}q^2$ = the probability of exactly $(m-2)$ elements operating successfully

$\frac{m(m-1)\cdots(m-i+1)}{i!}R^{m-i}q^i$ = the probability of exactly $(m-i)$ elements operating successfully

q^m = the failure probability of m elements

For nonidentical elements, the left-hand side of Eq. (9.60) becomes

$$\prod_{i=1}^{m}(p_i + q_i) \tag{9.61}$$

where

$$p_i = R_i$$

The series network reliability R_{sn} is given by the first right-hand term of Eq. (9.60):

$$R_{sn} = R^m \tag{9.62}$$

For nonidentical elements, the preceding expression becomes

$$R_{sn} = \prod_{i=1}^{m} R_i \tag{9.63}$$

where R_i is the ith element reliability; for $i = 1, 2, 3, \ldots, m$.

Example 9.14 An aircraft has three statistically independent engines. In other words, the engines fail independently. All three engines are required for the aircraft to fly successfully. The reliabilities of engines 1, 2, and 3 are 0.9, 0.95, and 0.92, respectively. Calculate the reliability of the aircraft flying successfully.

This problem is represented by the block diagram shown in Figure 9.2. With the aid of Eq. (9.63), the reliability of the series network shown in Fig. 9.2 is

$$R_{sn} = R_1 R_2 R_3 = (0.9)(0.95)(0.92)$$
$$= 0.7866$$

Thus the reliability of the aircraft flying successfully is 0.7866.

9.4.2 Parallel Model

This model is used to represent a system with m active and independent elements connected in parallel. That means that at least one element must function normally for the system's success. The system failure probability, F_{pn}, is given by the last right-hand term of Eq. (9.60) as

$$F_{pn} = q^m \tag{9.64}$$

where m is the number of elements and q is the failure probability of an element.

For nonidentical elements, the preceding equation becomes

$$F_{pn} = \prod_{i=1}^{m} q_i \tag{9.65}$$

The reliability of the network is given by

$$R_{pn} = 1 - F_{pn}$$

$$= 1 - \prod_{i=1}^{m} q_i \tag{9.66}$$

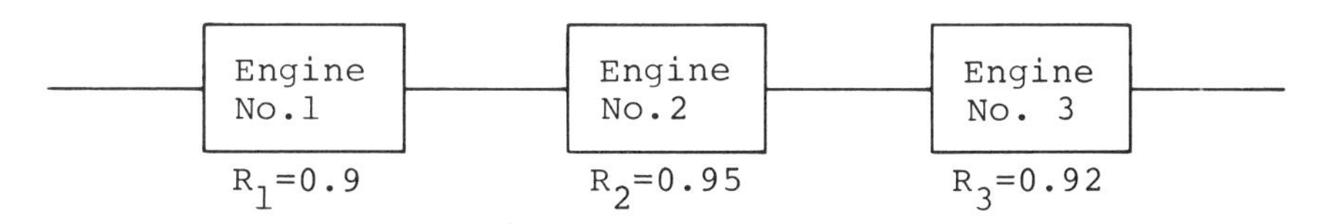

Figure 9.2 A series network.

Example 9.15 A system is composed of three independent and identical electric motors in parallel. At least one motor must function successfully for the system to operate normally. The failure probability of a motor is 0.15. Calculate the system reliability.

Using the specified data in Eq. (9.66) results in

$$R_{pn} = 1 - q^3 = 1 - (0.15)^3 = 0.99663$$

Thus the reliability of the system is 0.99663.

9.4.3 Partially Redundant System Model

This is another type of reliability model which is used to improve system reliability. Usually, it is known as the n-out-of-m units model. In other words, at least n units must function normally for the system's success instead of only one in the parallel case and all in the series case. The reliability expression for this model can be obtained from Eq. (9.60). Thus, the probability P_{en} of exactly n units operating successfully out of m units from Eq. (9.34) is

$$P_{en} = \binom{m}{n} R^n q^{m-n} \tag{9.67}$$

where R is the unit reliability and q is the unit failure probability.

Similarly from Eq. (9.60), the m-out-of-n units system reliability is given by

$$R_{n/m} = \sum_{i=m}^{m} \binom{m}{i} R^i (1 - R)^{m-i} \tag{9.68}$$

Example 9.16 An independent unit reliability network is shown in Fig. 9.3. The network is composed of three subsystems A, B, and C. In subsystem A only two units have to operate normally for the subsystem success. Each unit reliability is specified in Fig. 9.3. Calculate the network reliability.

By substituting the specified data into Eq. (9.68), the subsystem "A" is

$$\begin{aligned} R_A &= \sum_{i=2}^{3} \binom{3}{i} R^i (1 - R)^{3-i} \\ &= \binom{3}{2} R^2 (1 - R) + \binom{3}{3} R^3 \\ &= 3R^2 - 2R^3 = 3(0.8)^2 - 2(0.8)^3 = 0.896 \end{aligned}$$

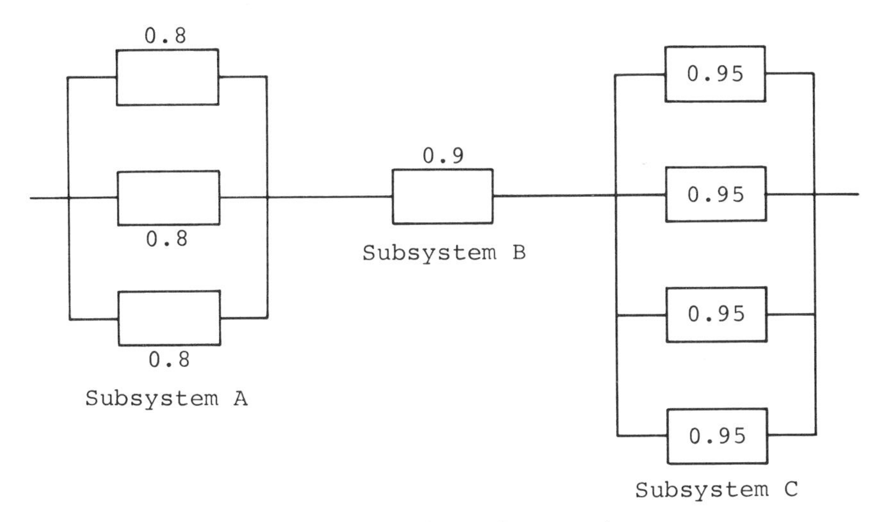

Figure 9.3 A complex network.

The subsystem "B" reliability is

$$R_B = 0.9$$

Finally, using the given data in Eq. (9.66), the subsystem "C" reliability is

$$\begin{aligned} R_C &= 1 - q^4 = 1 - (1 - R)^4 \\ &= 1 - (1 - 0.95)^4 \\ &= 0.999994 \end{aligned}$$

Thus the reliabilities of subsystems, A, B, and C are 0.896, 0.9, and 0.999994, respectively. That means we have reduced Fig. 9.3 to the network shown in Fig. 9.4.

With the aid of Eq. (9.63), the Fig. 9.4 network reliability is

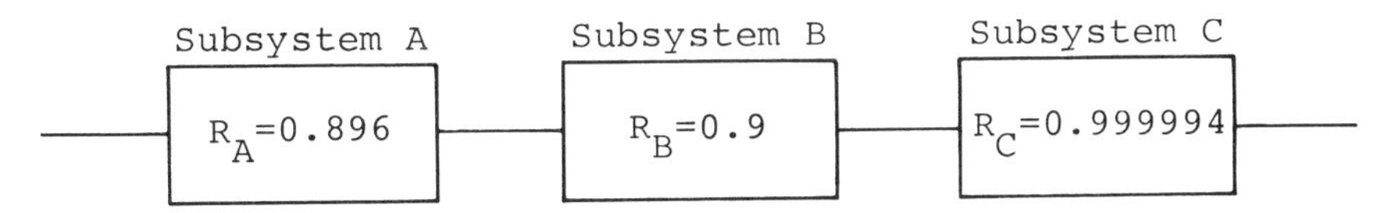

Figure 9.4 The reduced network.

$$
\begin{aligned}
R_N &= R_A R_B R_C \\
&= (0.896)(0.9)(0.999994) \\
&= 0.8064
\end{aligned}
$$

Thus, the network shown in Fig. 9.3 is 0.8064.

9.5 SUMMARY

This chapter briefly presents some introductory aspects of reliability. In broad terms the chapter begins by defining the cumulative distribution function, probability density function, reliability function, hazard rate, mean time to failure, and variance. The next topic covered in the chapter is probability distributions. Five such distributions are described: the binomial, Poisson, bathtub, exponential, and Weibull distributions.

The last topic of the chapter is reliability networks such as series, parallel, and n-out-of-m units. Equations to evaluate reliability of such networks are developed with the aid of the binomial formula. This chapter contains 16 examples along with their solutions.

EXERCISES

1. Prove that the reliability $R(t)$ of an item is given by

$$R(t) = \exp\left[-\int_0^t h(x)\,dx\right] \tag{9.70}$$

   where t is time and $h(t)$ is the hazard rate.
2. The probability density function $f(t)$ of a distribution is defined as

$$f(t) = \lambda t \exp\left(-\frac{\lambda t^2}{2}\right) \tag{9.71}$$

   where λ is the scale parameter and t is time. Obtain expressions for both reliability function and hazard rate. What is the name of this probability density function?
3. Prove that the expected value $E(t)$ of a failure density function is given by

$$E(t) = \lambda(\theta + 1) \tag{9.72}$$

   where λ is the scale parameter and θ is the shape parameter.
4. The failure rate of an electronic device is given by

$$\lambda = 0.003 \text{ failures/hr}$$

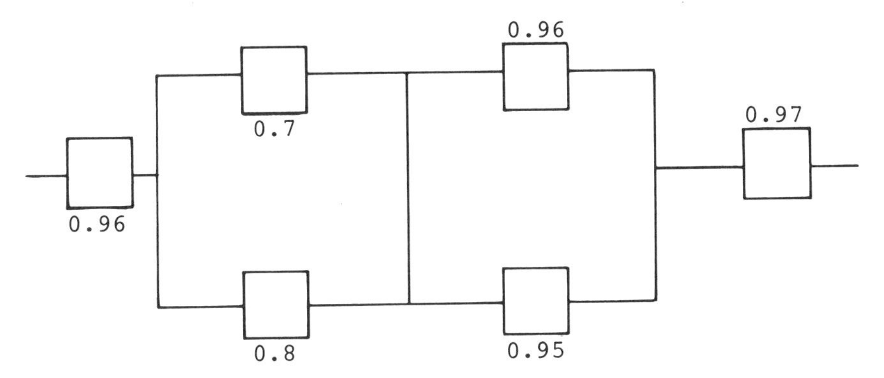

Figure 9.5 A reliability network.

The device is to be operated for a 100-hr mission. Calculate the failure probability of the device for the specified period.

5. Reliability of a three-independent-unit series system is given by

$$R(t) = \exp[-(\lambda_1 + \lambda_2 + \lambda_3)t \qquad (9.73)$$

where

λ_1 = the constant failure rate of unit 1
λ_2 = the constant failure rate of unit 2
λ_3 = the constant failure rate of unit 3
t = time

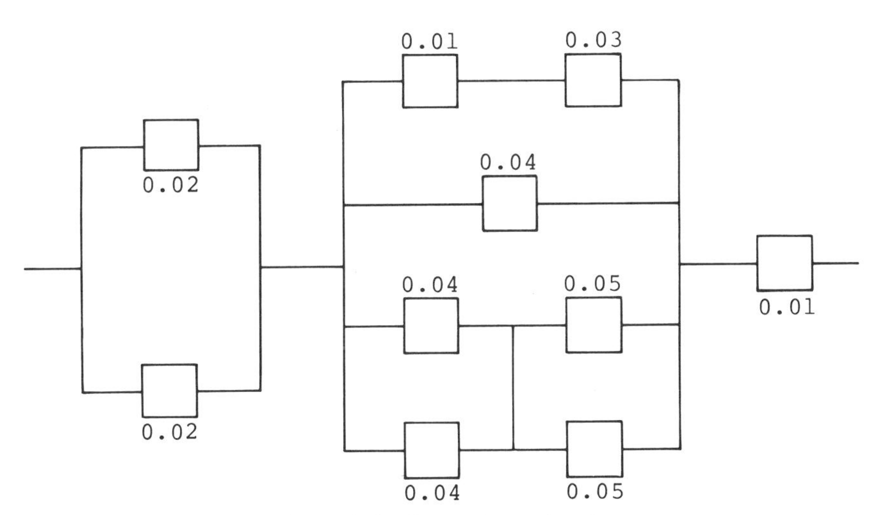

Figure 9.6 A complex network.

Failure rates of units 1, 2, and 3 are 0.0004, 0.002, and 0.0006 failures/hr, respectively. Calculate the system mean time to failure.

6. Calculate the failure probability of the independent unit network shown in Fig. 9.5. The reliability of each unit is specified in the diagram.
7. A complex network composed of independent units is shown in Fig. 9.6. The failure probability of each unit is specified in the diagram. Evaluate the reliability of the network.

REFERENCES

1. M. L. Shooman, *Probabilistic Reliability: An Engineering Approach*, McGraw-Hill, New York, 1968, p. 12.
2. Engineer Design Handbook, Development Guide for Reliability, AMCP 706-197, Published by U.S. Army Material Command, 1976. Available from 5001 Eisenhower Avenue, Alexandria, Virginia 22333.
3. D. K. Lloyd and M. Lipow, *Reliability: Management, Methods, and Mathematics,* Prentice-Hall, Englewood Cliffs, New Jersey, 1962, pp. 115.
4. W. Weibull, A Statistical Function of Wide Applicability, *J. Appl. Mech.*, Vol. 18 (1951), pp. 293–297.

10

Design for Reliability

10.1 INTRODUCTION

To produce reliable systems the reliability has to be considered at the system design stage. Otherwise, it may not be possible to have a high-reliability product in the field. At the design stage, the system under consideration is carefully examined from different angles. The weak areas of the system from the reliability point of view are identified by performing reliability analysis. In addition, the weakness in the system is overcome by introducing redundancy or using more reliable system components or modifying the design.

Therefore, this chapter briefly presents the various aspects of reliability—for example, failure data analysis, reliability allocation, reliability analysis techniques, failure rate evaluation of electronic components, and systems and software reliability.

10.2 FAILURE DATA ANALYSIS: THE MAXIMUM LIKELIHOOD ESTIMATION METHOD

Once the failure data on a component, a device, or a system are available, then the next logical step is to analyze such data. This basically involves ascertaining the type of probability density function which will best fit the given failure data and estimating the values of the distribution parameters. Therefore, this section presents the maximum likelihood method to obtain point estimates for continuous distributions. Other failure data analysis procedures may be found in Refs. 1–3.

The likelihood function L is defined as follows:

$$L(t_1, t_2, t_3, \ldots, t_k; \alpha_1, \alpha_2, \alpha_3, \ldots, \alpha_n) = f(t_1; \alpha_1, \alpha_2, \alpha_3, \ldots, \alpha_n) \times f(t_2; \alpha_1, \alpha_2, \alpha_3, \ldots, \alpha_n) \cdots f(t_k; \alpha_1, \alpha_2, \alpha_n) \quad (10.1)$$

where

k = the number of data points
n = the number of parameters to be estimated
α_i = the ith parameter, for $i = 1, 2, 3, n$
t_i = the ith time of failure, for $i = 1, 2, 3, k$
$f(t_i; \alpha_1, \alpha_2, \alpha_3, \ldots \alpha_n)$ = the failure probability density function at the time of failure t_i; for $i = 1, 2, 3, \ldots, n$

Taking the logarithms of Eq. (10.1), we get

$$\ln L(t_1, t_2, t_3, \ldots, t_k; \alpha_1, \alpha_2, \alpha_3, \ldots, \alpha_n) = \ln f(t_1; \alpha_1, \alpha_2, \alpha_3, \ldots, \alpha_n) + \ln f(t_2; \alpha_1 \alpha_2, \alpha_3, \ldots, \alpha_n) + \cdots + \ln f(t_k; \alpha_1, \alpha_2, \alpha_3, \ldots, \alpha_n) \quad (10.2)$$

Taking the partial derivative of Eq. (10.2) with respect to α_i and then equating to zero yields

$$\frac{\partial \ln L(t_1, t_2, t_3, \ldots, t_k; \alpha_1, \alpha_2, \alpha_3, \ldots, \alpha_n)}{\partial \alpha_i} = 0 \quad \text{for } i = 1, 2, 3, \ldots, n \quad (10.3)$$

The values for α_i, for $i = 1, 2, 3, \ldots, n$, are estimated by solving the set of Eqs. (10.3).

For large k [4], the variance of $\hat{\alpha}_i$ is given by

$$\operatorname{var} \hat{\alpha}_i = -\left[\frac{\partial^2 \ln L(\cdot;\cdot)}{\partial \alpha_i^2}\right]^{-1} \quad \text{for } i = 1, 2, 3, \ldots, n \quad (10.4)$$

Since the exponential distribution is frequently used in reliability work, the maximum likelihood estimation method is demonstrated by using such a distribution in the following examples.

Example 10.1 Failure times of an electronic device are defined by the following probability density function:

$$f(t) = \frac{1}{\mu} \exp\left(-\frac{1}{\mu} t\right) \quad (10.5)$$

where $f(t)$ is the failure probability density function, t is time, and μ is the mean of the exponential distribution. Derive an expression for $\hat{\mu}$ with the aid of the maximum likelihood estimation method. The symbol $\hat{\mu}$ signifies the maximum likelihood estimate of μ.

Substituting Eq. (10.5) into Eq. (10.2) yields

$$\ln L(t;\mu) = \ln\left[\frac{1}{\mu}\exp\left(-\frac{1}{\mu}t_1\right)\right] + \ln\left[\frac{1}{\mu}\exp\left(-\frac{1}{\mu}t_2\right)\right] + \cdots + \ln\left[\frac{1}{\mu}\exp\left(-\frac{1}{\mu}t_k\right)\right] \tag{10.6}$$

$$\ln L(t;\mu) = k \ln\left(\frac{1}{\mu}\right) - \frac{1}{\mu}\sum_{i=1}^{k} t_i \tag{10.7}$$

Differentiating Eq. (10.7) with respect to μ results in

$$\frac{\partial \ln L(t;\mu)}{\partial \mu} = -\frac{k}{\mu} + \frac{1}{\mu^2}\sum_{i=1}^{k} t_i \tag{10.8}$$

Setting the above equation equal to zero yields

$$-\frac{k}{\mu} + \frac{1}{\mu^2}\sum_{i=1}^{k} t_i = 0 \tag{10.9}$$

Rearranging Eq. (10.9) leads to

$$\mu = \frac{1}{k}\sum_{i=1}^{k} t_i \tag{10.10}$$

Example 10.2 Failure times of electric bulbs are described by an exponential distribution. Over a period of time 12 identical electric bulbs were observed. They failed after 200, 100, 250, 300, 400, 200, 350, 50, 250, 300, 500, and 300 hr of use, respectively. Estimate the value of the mean life of the bulbs.

In this example the following data are specified for the components of Eq. (10.10):

$k = 12$, $t_1 = 200$ hr, $t_2 = 100$ hr, $t_3 = 250$ hr, $t_4 = 300$ hr, $t_5 = 400$ hr, $t_6 = 200$ hr, $t_7 = 350$ hr, $t_8 = 50$ hr, $t_9 = 250$ hr, $t_{10} = 300$ hr, $t_{11} = 500$ hr, $t_{12} = 300$ hr

Substituting the above data into Eq. (10.10) results in

$$\hat{\mu} = \frac{1}{12}(200 + 100 + 250 + 300 + 400 + 200 + 350 + 50$$
$$+ 250 + 300 + 500 + 300)$$
$$= 266.67 \text{ hr}$$

Thus the mean life of an electric bulb is 266.67 hr

Example 10.3 For the failure probability density function given as Eq. (10.5), obtain an expression for the variance of $\hat{\mu}$. Assume that there are a large number of data points.

Differentiating Eq. (10.8) with respect to μ, we get

$$\frac{\partial^2 \ln L(t,\mu)}{\partial \mu^2} = +\frac{k}{\mu^2} - \frac{2}{\mu^3} \sum_{i=1}^{k} t_i \tag{10.11}$$

From Eq. (10.10), we get

$$\sum_{i=1}^{k} t_i = \mu k \tag{10.12}$$

Thus substituting Eq. (10.12) into Eq. (10.11) leads to

$$\frac{\partial^2 \ln L(t,\mu)}{\partial \mu^2} = \frac{k}{\mu^2} - \frac{2}{\mu^3} \mu^k = -\frac{k}{\mu^2} \tag{10.13}$$

Substituting Eq. (10.13) into Eq. (10.4) leads to

$$\text{var}\, \hat{\mu} = \frac{\mu^2}{k} \tag{10.14}$$

10.3 RELIABILITY ALLOCATION

Once the overall system reliability quantitative requirements are specified, the next logical step is to allocate the specified reliability to its subsystems or components at the system developmental stage. This way if the subsystems' or components' allocated reliability requirements are met then the overall specified system reliability will automatically be achieved. According to Ref. 5, it may be said that the reliability allocation is the process of assigning reliability requirements to system parts or components to achieve the overall specified reliability of a system. The main benefits of the reliability allocation are as follows:

1. It forces the designer to consider reliability on an equal basis with other system parameters.
2. It helps to force the designer to comprehend and establish relationships among the reliabilities of component, subsystem, and system. In this way, in design, it helps one to understand the fundamental reliability problems.
3. It leads to better design, manufacturing methods, etc.

Various different reliability procedures are used to allocate specified reliability. Their application depends upon one's needs. However, here we will present two such allocation procedures. Many more procedures are given in Refs. 5 and 6.

10.3.1 The ARINC Method

This technique was developed by ARINC Research Corporation [7] to allocate specified system reliability. The following assumptions are associated with the technique:

1. The system is composed of subsystems forming a series configuration. In other words, if any one of the subsystems fails, the entire system will fail.
2. Failure rates of subsystems are constant.
3. Subsystem failures are independent.
4. Subsystem mission times are equal to the system mission time.

Thus a series system reliability, $R_s(t)$, is given by

$$R_s(t) = \prod_{i=1}^{m} R_i(t) \tag{10.15}$$

where $R_i(t)$ is the ith subsystem reliability at time t and m is the number of subsystems.

For the constant failure rate of the ith subsystem

$$R_i(t) = e^{-\lambda_i t} \tag{10.16}$$

where λ_i is the constant failure rate of the ith subsystem.

Similarly for the constant failure rate, the series system reliability, $R_s(t)$, is given by

$$R_s(t) = e^{-\lambda_{cs} t} \tag{10.17}$$

where λ_{cs} is the constant failure rate of the series system.

Substituting Eqs. (10.16)–(10.17) into Eq. (10.15) leads to

$$e^{-\lambda_{cs}t} = \prod_{i=1}^{m} e^{-\lambda_i t} = \exp\left(-\sum_{i=1}^{m} \lambda_i t\right) \tag{10.18}$$

Taking the natural logarithms of both sides of Eq. (10.18), we get

$$\lambda_{cs'} = \sum_{i=1}^{m} \lambda_i \tag{10.19}$$

Thus we have to choose $\hat{\lambda}_i$, the allocated failure rates of the subsystems, in such a way that

$$\sum_{i=1}^{m} \hat{\lambda}_i \leq \lambda \tag{10.20}$$

where λ is the constant failure rate specified for the system.

Now, the allocation process can be accomplished in the following three steps:

1. Obtain the failure rate estimate for each subsystem from the field data or other sources (i.e., estimate for $\lambda_1, \lambda_2, \lambda_3, \ldots, \lambda_m$).
2. Calculate the relative subsystem weights. For example, the ith subsystem weighting factor is given by

$$\beta_{wi} = \lambda_i / \left(\sum_{i=1}^{m} \lambda_i\right) \qquad \text{for } i = 1, 2, 3, \ldots, m \tag{10.21}$$

   One should note here that $\Sigma_{i=1}^{m} \lambda_{wi} = 1$, because β_{wi} denotes the relative failure vulnerability of the subsystem i.
3. Calculate the subsystem i failure rate allocation from the following equation:

$$\hat{\lambda}_i = \beta_{wi} \lambda \tag{10.22}$$

Example 10.4 A series system is composed of five independent subsystems. The estimated failure rates for subsystems 1, 2, 3, 4, and 5 are $\lambda_1 = 0.0001$, $\lambda_2 = 0.0002$, $\lambda_3 = 0.0003$, $\lambda_4 = 0.0004$, and $\lambda_5 = 0.0005$ failures/hr, respectively. The specified failure rate of the series system is $\lambda = 0.002$ failures/hr. Calculate the value of the failure rate to be allocated to each subsystem.

Substituting the specified data into Eq. (10.19) results in

$$\lambda_{cs} = \sum_{i=1}^{5} \lambda_i = \lambda_1 + \lambda_2 + \lambda_3 + \lambda_4 + \lambda_5$$

$$= 0.0001 + 0.0002 + 0.0003 + 0.0004 + 0.0005$$
$$= 0.0015 \text{ failures/hr}$$

Thus, from Eq. (10.21), the subsystems' 1, 2, 3, 4, and 5 weights, respectively, are

$$\beta_{w1} = \frac{\lambda_1}{\lambda_{cs}} = \frac{0.0001}{0.0015} = 0.0667$$

$$\beta_{w2} = \frac{\lambda_2}{\lambda_{cs}} = \frac{0.0002}{0.0015} = 0.1333$$

$$\beta_{w3} = \frac{\lambda_3}{\lambda_{cs}} = \frac{0.0003}{0.0015} = 0.2$$

$$\beta_{w4} = \frac{\lambda_4}{\lambda_{cs}} = \frac{0.0004}{0.0015} = 0.2667$$

$$\beta_{w5} = \frac{\lambda_5}{\lambda_{cs}} = \frac{0.0005}{0.0015} = 0.3333$$

Finally, for subsystems, 1, 2, 3, 4, and 5, respectively, the allocated failure rates are as follows:

$$\hat{\lambda}_1 = \beta_{w1}\lambda = (0.0667)(0.002) = 0.0001 \text{ failures/hr}$$
$$\hat{\lambda}_2 = \beta_{w2}\lambda = (0.1333)(0.002) = 0.0003 \text{ failures/hr}$$
$$\hat{\lambda}_3 = \beta_{w3}\lambda = (0.2)(.002) = 0.0004 \text{ failures/hr}$$
$$\hat{\lambda}_4 = \beta_{w4}\lambda = (0.2667)(0.002) = 0.0005 \text{ failures/hr}$$
$$\hat{\lambda}_5 = \beta_{w5}\lambda = (0.3333)(0.002) = 0.0007 \text{ failures/hr}$$

10.3.2 Repairable Systems Allocation Method

This method is used to allocate failure rate and repair rate for systems composed of subsystems in series [5]. The following assumptions are associated with this technique:

1. The system is composed of k independent subsystems in series.
2. Subsystems are identical.
3. Each subsystem is assigned to one repairman; in other words, $\mu/\lambda \gg 1$ (where λ is the subsystem constant failure rate and μ is the subsystem constant repair rate).

The series system steady-state availability AV_{ss} is given by

$$AV_{ss} = \prod_{i=1}^{k} A_{si} \tag{10.23}$$

where k is the number of subsystems in the series and A_{si} is the ith subsystem's steady-state availability.

The subsystem i steady-state availability from Ref. 4 is given by

$$A_{si} = \frac{\mu_i}{\mu_i + \lambda_i} \tag{10.24}$$

where μ_i is the ith subsystem constant repair rate and λ_i is the ith subsystem constant failure rate.

Thus, for identical subsystems, substituting Eq. (10.24) into Eq. (10.23) yields

$$AV_{ss} = A_s^k = \left(\frac{\mu}{\mu + \lambda}\right)^k$$

$$= \left(\frac{1}{1 + \theta}\right)^k \tag{10.25}$$

where

$$\theta = \frac{\lambda}{\mu}$$

The allocation procedure is outlined in the following steps:

1. Specify the series system overall steady-state availability.
2. Determine the steady-state availability of each subsystem by utilizing the following equation [this equation is obtained by utilizing the left-hand side of Eq. (10.25)]:

$$A_s = (AV_{ss})^{1/k} \tag{10.26}$$

3. Determine the value of θ for a subsystem: This can be easily obtained by utilizing Eq. (10.24). For example, a single subsystem availability is given by

$$A_s = \frac{\mu}{\mu + \lambda} = \frac{1}{1 + \lambda/\mu} = \frac{1}{1 + \theta} \tag{10.27}$$

Rearranging Eq. (10.27), we get

$$\theta = \left(\frac{1}{A_s} - 1\right) \tag{10.28}$$

4. Calculate the value to be allocated to each subsystem. For example, the value to be allocated for each subsystem's failure rate $\hat{\lambda}$ is given by

$$\hat{\lambda} = \theta\mu \tag{10.29}$$

where μ is the specified value for the repair rate of a subsystem.

Example 10.5 A series system is composed of seven statistically independent subsystems. All the subsystems are identical. Each subsystem will be assigned to one repairman. The specified system availability and the repair rate are 0.98 and 0.25 repairs/hr, respectively. Calculate the value of failure rate to be allocated to each subsystem.

By substituting the specified data into Eq. (10.26) we get

$$A_s = (0.98)^{1/7} = 0.99712$$

Substituting the above value into Eq. (10.28) leads to

$$\theta = \left(\frac{1}{0.99712} - 1\right) = 0.002888$$

Thus utilizing the above value and the specified repair rate data in Eq. (10.29) results in

$$\begin{aligned}\hat{\lambda} &= (0.002888)(0.25) \\ &= 0.000722 \text{ failures/hr}\end{aligned}$$

Each subsystem will have the allocated failure rate of 0.000722 failures/hr.

10.4 RELIABILITY ANALYSIS TECHNIQUES

This section briefly presents the two commonly used methods to perform reliability analysis at the system developmental stage. These are the failure modes and effects analysis (FMEA) and the fault tree analysis. Both these techniques are described briefly below:

10.4.1 Failure Modes and Effects Analysis

This technique was first used in the development of flight control systems in the early 1950s [8]. The FMEA method is concerned with determining design reliability by considering potential failures and their effect on the system under study. More clearly, it is concerned with listing each failure mode of an item on paper and its effects on the listed subsystems. This "bottom-up" approach can be utilized at any level, that is, from complete systems to components. In broader terms, the following steps are involved in performing failure modes and effects analysis:

1. Define the boundaries of the system in question and its requirements.
2. Establish ground rules to perform FMEA.
3. List the system's subsystems and parts.
4. List for the part under consideration its identification, failure modes, and description.
5. For each failure mode estimate its failure rate.
6. List the effects of each failure mode on subsystem, system, and so on.
7. For each failure mode enter remarks.
8. Investigate the serious failure modes and take desirable steps.

Some of the advantages of the FMEA are as follows:

1. It is a visibility tool; that is, FMEA tabulations are visible.
2. It is a systematic procedure.
3. It can easily be understood.
4. It identifies weaknesses in the system design.
5. It becomes useful in design comparison.

10.4.2 Fault Tree Technique

This is a commonly used technique in industry to evaluate reliability of a system. It was developed in the early 1960s to evaluate reliability of the Minuteman Launch Control System. Since then it has gained favor over other techniques especially when analyzing complex systems. Fault tree analysis begins by identifying the top event, known as the undesirable event of the system. The undesirable event of the system is caused by events generated and connected by logic gates such as AND, OR, etc. The following basic steps are involved in performing fault tree analysis:

1. Establishing system definition.
2. Constructing the fault tree.
3. Evaluating the fault tree qualitatively.
4. Collecting basic data such as components' failure rates, repair rates, and failure occurrence probability.
5. Evaluating the fault tree quantitatively.
6. Recommending corrective measures.

Some of the fault tree symbols are given in Fig. 10.1.

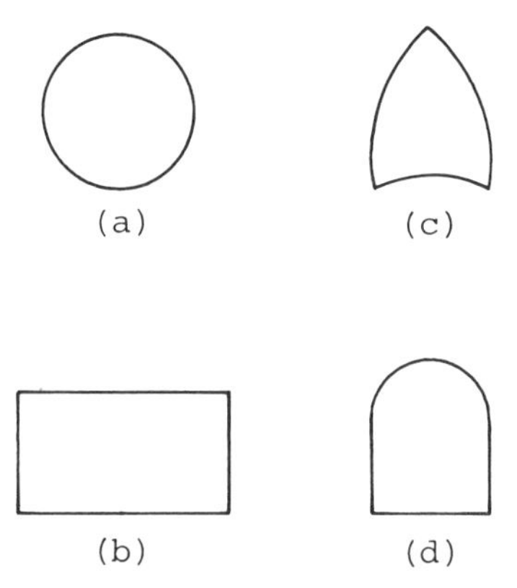

Figure 10.1 Basic fault tree symbols and their definitions. (a) Basic fault event (e.g., failure of an elementary component) which can be assigned a probability of occurrence, a failure rate, and a repair rate. (b) Fault event resulting from the combination of fault events through the input of a logic gate such as OR and AND. (c) OR logic gate. This gate signifies that the output fault event occurs when one or more of the input events occur. (d) AND gate. This gate signifies that the output fault event occurs only when all the input fault events occur.

An example of a fault tree is shown in Fig. 10.2. This fault tree is concerned with a room containing three light bulbs controlled from a single switch. In this case, the "room without electric light" is the "undesired event." In other words, this is the event we wish to investigate. The undesired event can occur due to the occurrence of any one of three input events "bulbs burnt out," "switch fails to close," and "no electricity." Thus this situation is represented by a three-input OR gate. The occurrence of events "bulbs burnt out" and "no electricity" is investigated further. The event "bulbs burnt out" occurs when all three bulbs are burnt out. Thus this situation is represented by a three-input AND gate. The event "no electricity" occurs when power is cut off or the fuse is blown.

Probability Evaluation

This section is concerned with the quantitative evaluation of fault trees. In addition, two different situations are considered:

1. Logic gate input fault events (e.g., component failures) are nonrepairable.
2. Logic gate input fault events are repairable.

For the first situation, the following formulas [9] are associated with OR and AND gates:

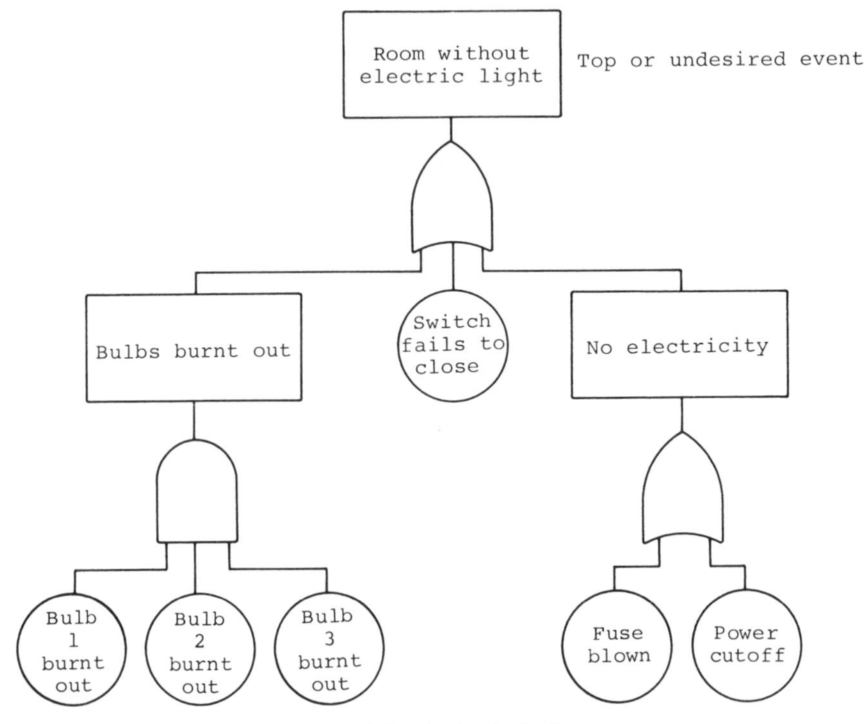

Figure 10.2 A simple fault tree.

OR gate

$$E_0 = E_1 + E_2 + E_3 + \cdots + E_m \tag{10.30}$$

where E_0 is the output event of the OR gate; E_i is the ith input event of the OR gate, for $i = 1, 2, \ldots, m$; and m is the number of input fault events.

Thus the probability of the occurrence of event E_0 is given by

$$P(E_0) = 1 - \prod_{i=1}^{m} [1 - P(E_i)] \tag{10.31}$$

where $P(E_0)$ is the probability of occurrence of output event E_0 and $P(E_i)$ is the probability of occurrence of the independent input fault event E_i, for $i = 1, 2, 3, \ldots, m$.

AND gate

$$E_{0A} = E_1 E_2 E_3 \cdots E_m \tag{10.32}$$

where E_{0A} is the output event of the AND gate and, E_i is the ith input event of the AND gate; for $i = 1, 2, 3, \ldots, m$.

Thus the probability of occurrence of event E_{0A} is given by

$$P(E_{0A}) = \prod_{i=1}^{m} P(E_i) \tag{10.33}$$

where $P(E_{0A})$ is the probability of occurrence of output event E_{0A} and $P(E_i)$ is the probability of occurrence of the ith independent input fault event E_i, for $i = 1, 2, 3, \ldots, m$.

Similarly for situation 2 (i.e., logic gate input fault events are repairable), the following formulas are associated with OR and AND gates.

OR gate

$$U_0 = 1 - \prod_{i=1}^{m} \{1 - U_i\} \tag{10.34}$$

where U_0 is the unavailability (probability of occurrence) associated with the output event of the OR gate and U_i is the steady-state unavailability (probability of occurrence) associated with the ith independent input fault event, for $i = 1, 2, \ldots, m$.

The steady-state unavailability U_i from Ref. 9 is given by

$$U_i = \frac{\lambda_i}{\lambda_i + \mu_i} \tag{10.35}$$

where λ_i is the constant failure rate of the ith component (i.e., fault event occurrence rate), for $i = 1, 2, 3, \ldots, m$ and μ_i is the constant repair rate of the ith component (i.e., fault event), for $i = 1, 2, 3, \ldots, m$.

AND gate

$$U_{0A} = \prod_{i=1}^{m} U_i \tag{10.36}$$

where U_{0A} is the unavailability (probability of occurrence) associated with the output event of the AND gate.

Example 10.6 A four-independent-input OR gate is shown in Fig. 10.3. Each input fault event occurrence probability is specified. Calculate the output event probability of occurrence.

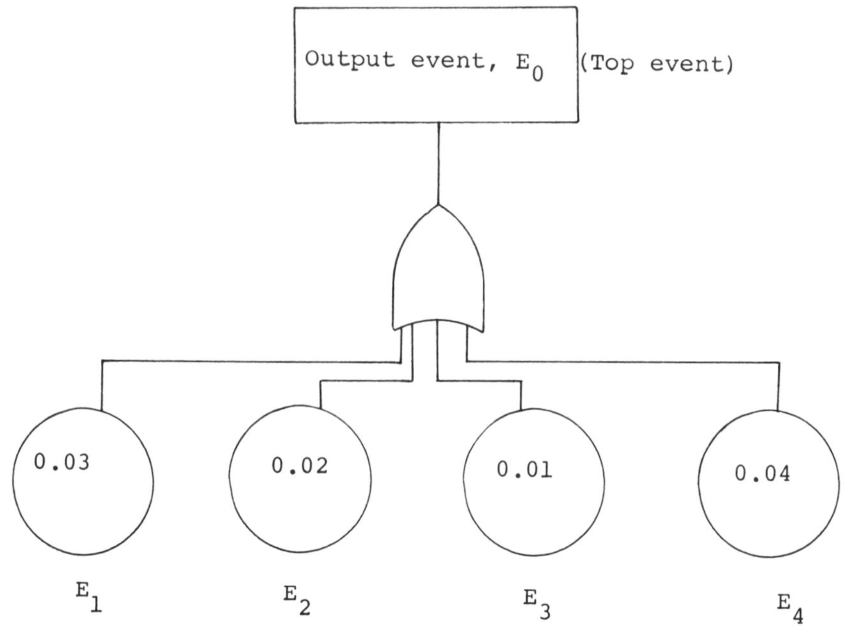

Figure 10.3 A four-input OR gate.

From Eq. (10.31) for the specified data we get

$$\begin{aligned} P(E_0) &= 1 - [1 - P(E_1)][1 - P(E_2)][1 - P(E_3)][1 - P(E_4)] \\ &= 1 - (1 - 0.03)(1 - 0.02)(1 - 0.01)(1 - 0.04) \\ &= 0.0965 \end{aligned}$$

Thus the probability of the top event occurrence is 0.0965.

Example 10.7 A three-independent-input AND gate is shown in Fig. 10.4. Input fault events 1, 2, and 3 constant occurrence rates are 0.002, 0.004, and 0.001 failures/hr, respectively. Similarly, the input fault events 1, 2, and 3 constant repair rates are 0.003, 0.005, and 0.002 repairs/hr, respectively.

1. Calculate the steady-state unavailability (occurrence probability) of each input fault event;
2. Calculate the probability of occurrence of the AND gate output event.

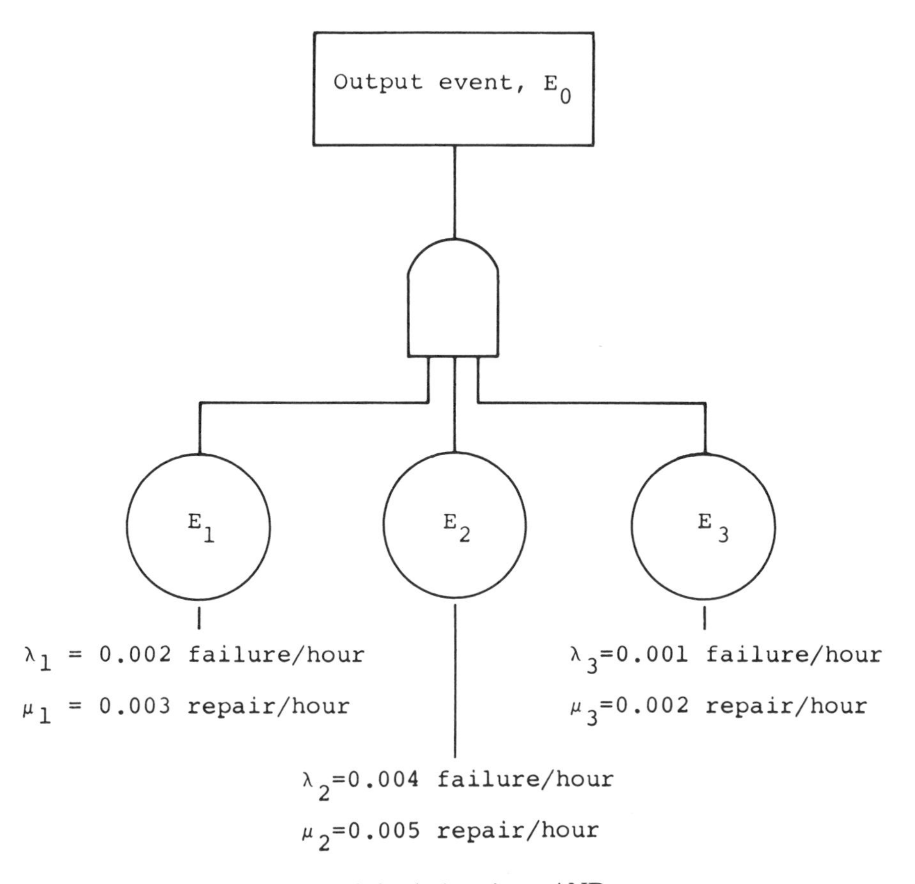

Figure 10.4 A three-input AND gate.

By substituting the specified data into Eq. (10.35) we get the unavailabilities (occurrence probabilities) of input fault events, 1, 2, and 3, respectively as follows:

$$U_1 = \frac{\lambda_1}{\lambda_1 + \mu_1} = \frac{0.002}{0.002 + 0.003} = 0.4$$

$$U_2 = \frac{\lambda_2}{\lambda_2 + \mu_2} = \frac{0.004}{0.004 + 0.005} = 0.444$$

and

$$U_3 = \frac{\lambda_3}{\lambda_3 + \mu_3} = \frac{0.001}{0.001 + 0.002} = 0.333$$

Thus substituting the above calculations into Eq. (10.36) results in

$$U_{0A} = U_1 U_2 U_3 = (0.4)(0.444)(0.333)$$
$$= 0.0591$$

Thus, the occurrence probability of the output event is 0.0591.

Example 10.8 The independent inputs and redundancy free fault tree shown in Fig. 10.2 is redrawn in Fig. 10.5. In Fig. 10.5 the constant failure and repair rates data for basic events are specified. Calculate the steady-state unavailability (probability of occurrence) of the top event.

Utilizing the specified data given in Fig. 10.5 in Eq. (10.35), the steady-state unavailabilities (probabilities of occurrence) associated with fault events "bulb burnt out," "switch fails to close," "fuse blown and power cut off," respectively, are as follows:

$$U = \frac{\lambda}{\lambda + \mu} = \frac{0.0004}{0.0004 + 0.0005} = 0.4444$$

$$U_1 = \frac{\lambda_1}{\lambda_1 + \mu_1} = \frac{0.0001}{0.0001 + 0.0004} = 0.0244$$

$$U_2 = \frac{\lambda_2}{\lambda_2 + \mu_2} = \frac{0.0003}{0.0003 + 0.0005} = 0.0566$$

and

$$U_3 = \frac{\lambda_3}{\lambda_3 + \mu_3} = \frac{0.0002}{0.0002 + 0.0005} = 0.0385$$

Utilizing one of the above calculations in Eq. (10.36), the occurrence probability of event A (shown in Fig. 10.5) is

$$U_A = U^3 = (0.4444)^3 = 0.0878$$

Similarly, substituting the calculated values of U_1 and U_2 into Eq. (10.34), the occurrence probability of event B (shown in Fig. 10.6) is

$$U_B = 1 - (1 - U_2)(1 - U_3)$$
$$= 1 - [1 - (0.0566)][1 - (0.0385)]$$
$$= 0.0929$$

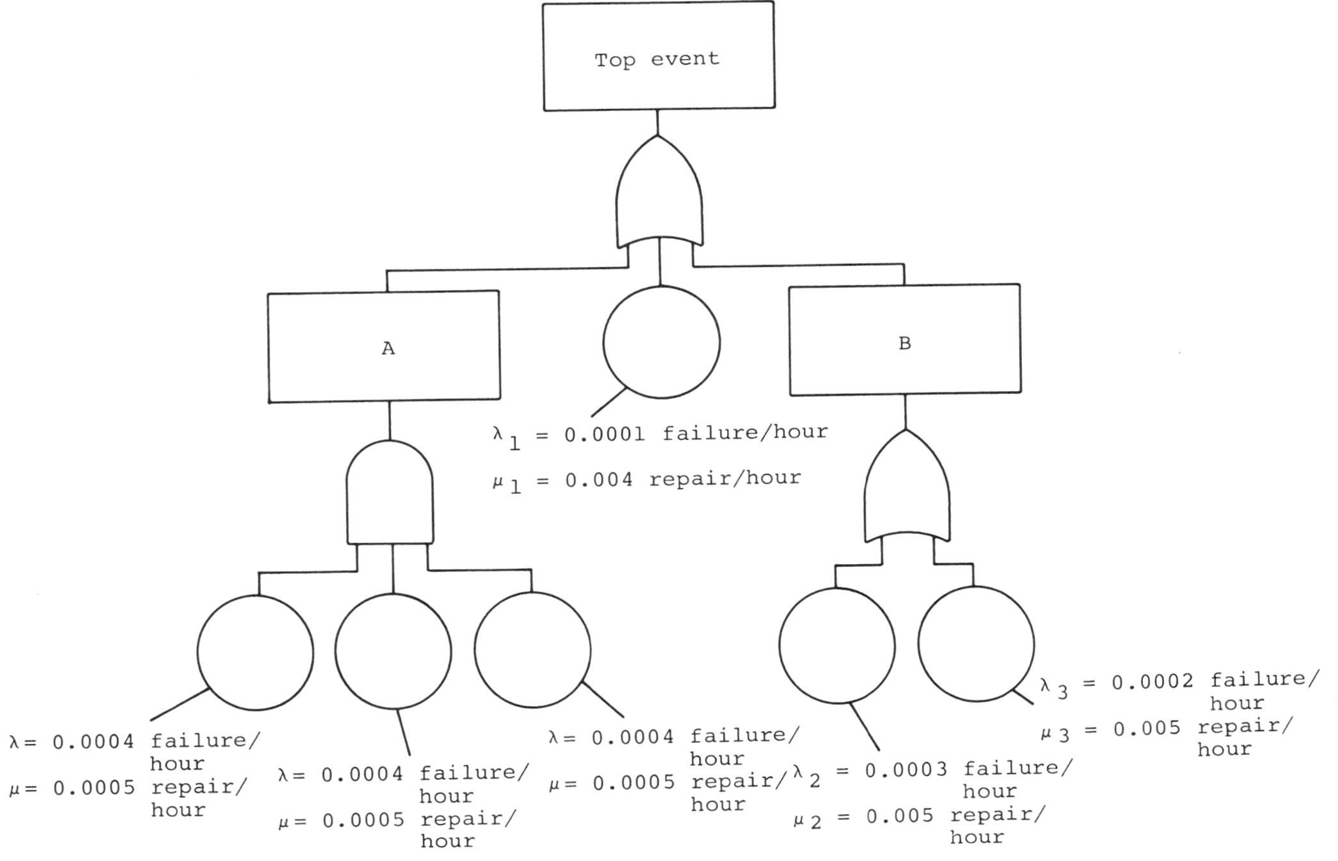

Figure 10.5 A fault tree with specified failure and repair rates data.

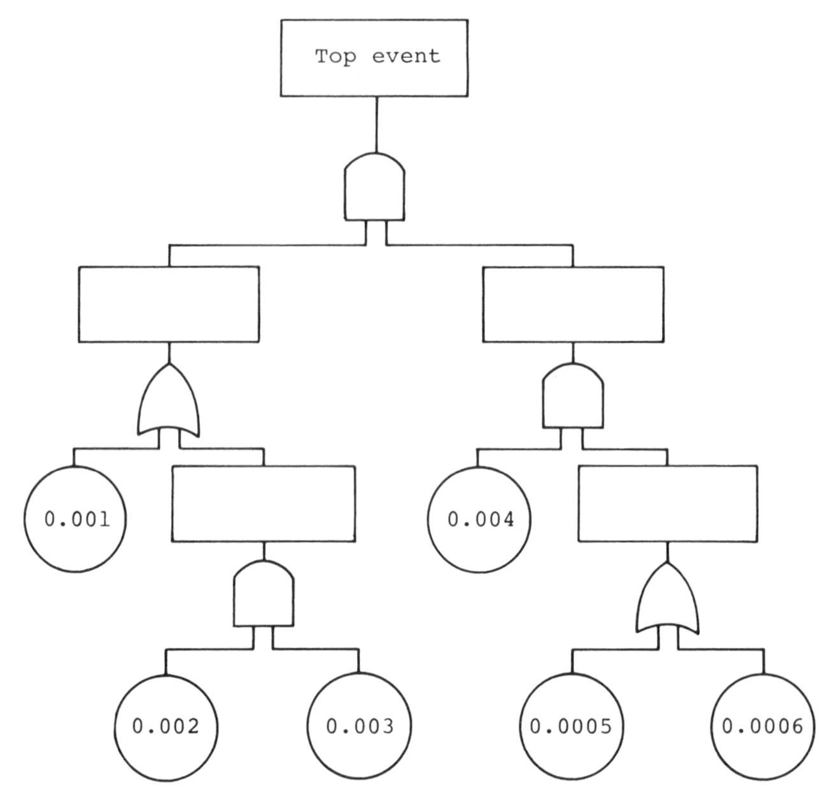

Figure 10.6 A fault tree with specified occurrence probabilities of basic fault events.

Finally, the probability of occurrence of the top event from Eq. (10.34) is

$$\begin{aligned} U_T &= 1 - (1 - U_A)(1 - U_B)(1 - U_1) \\ &= 1 - \{1 - (0.0878)\}[1 - (0.0929)][1 - (0.0244)] \\ &= 0.1927 \end{aligned}$$

Thus the probability of occurrence of the top event is 0.1927.

10.4.3 Conceptually Desirable Approach Open to Designer for Achieving Product Reliability

This approach is known as the "avoiding failure" approach [9]. It can be accomplished through the following:

1. Keeping the product under consideration as simple as possible

2. Considering the use of redundant items, whenever it is desirable
3. Testing the product design for its conformance with quality specification in its field environment
4. Considering the systems-design approach
5. Paying critical attention to reliability assessment both in production and in field
6. Studying the effect on reliability of innovation
7. Controlling the duty of product under design (whenever possible a product should be designed so that more than a safe load on the product cannot be applied)
8. Determining the overall reliability of the product under consideration
9. Studying the stress level on components
10. Giving full consideration to items that seem to be familiar and simple
11. Giving full attention to human limitations when designing a product
12. Examining variable interfaces between product parts

10.5 FAILURE RATE EVALUATION OF ELECTRONIC COMPONENTS AND SYSTEMS

This section describes models to evaluate failure rates of electronic parts and systems. These are as follows.

10.5.1 Failure Rate Evaluation of an Electronic Component

Many electronic components' base failure rate λ_b is based on the following equation:

$$\lambda_b = C \exp\left(\frac{-e_a}{kt}\right) \tag{10.37}$$

where

t = the absolute temperature
k = Boltzmann's constant
e_a = the activation energy for the process
C = a constant

Equation (10.37) is known as the Arrhenius model. In Ref. 10 to predict failure rates of a large number of electronic components, an equation of the following form is used:

$$\lambda_c = \lambda_B F_q F_e \cdots \tag{10.38}$$

where

λ_c = the constant failure rate of a component
F_q = the factor which takes into consideration part quality level
F_e = the factor which takes into consideration the influence of environment
λ_B = the base failure rate related to temperature and electrical stresses

The values of λ_B, F_q, F_e, etc., are specified for various different situations and components in Ref. 10.

Example 10.9 A tunnel diode failure rate λ_c from Ref. 10 is given by

$$\lambda_c = \lambda_B F_e F_q \text{ failures}/10^6 \text{ hr} \tag{10.39}$$

The following data are specified for the diode:

1. F_q (quality level − JAN) = 5.0
2. F_e (environment − ground, benign) = 1.0
3. λ_B (base failure rate − failures per 10^6 hours) = 0.044 failures/10^6 hr

The preceding data are taken from Ref. 10 for known factors associated with the diode, for example, quality level, usage environment, temperature, and stress ratio. Calculate the tunnel diode failure rate.

Substituting the specified data into Eq. (10.39) results in

$$\begin{aligned}\lambda_c &= (0.044)(1)(5)\\ &= 0.22 \text{ failures}/10^6 \text{ hr}\end{aligned}$$

Thus the tunnel diode failure rate is 0.22 failures/10^6 hr.

10.5.2 Failure Rate Evaluation of Equipment

This section presents a method known as the parts count method [10]. This method is used to predict failure rate of a system during early design stages. Furthermore, it also becomes useful during bid proposal. The following equation is used to calculate the constant failure rate λ_t of a piece of equipment:

$$\lambda_t = \sum_{j=1}^{k} q_j (\lambda_g F_q)_j \text{ failures}/10^6 \text{ hr} \tag{10.40}$$

where

k = the number of different generic component classifications
q_j = the jth generic component quantity
F_q = the jth generic component quality factor
λ_g = the constant generic failure rate of the jth generic component; this must be given in failures/10^6 hr

The values for F_q and λ_g are tabulated in Ref. 10. Equation (10.40) is subject to the assumption that the entire equipment or system will be operated under the same environment.

10.6 USEFUL GUIDELINES FOR EASE OF MAINTENANCE, PART SELECTION, AND CONTROL

This section briefly presents the design-related guidelines for both part selection and control, and for ease of maintenance. These are presented below, separately.

10.6.1 Useful Ease of Maintenance Design Guidelines

This section presents some of the ease of maintenance guidelines [11]. These are as follows:

1. Minimize special tool requirements.
2. Give full consideration when designing to ease of entry access. In other words, design in access which will simplify entry for maintenance.
3. Take advantage of modular design methods.
4. Design so that the diagnostic strategies are minimized.
5. Design to reduce interconnections of each replaceable part.
6. Enhance the technical maintenance literature quality.
7. Make use of special built-in fault detection circuits.
8. Minimize the isolation access need.
9. Make use of plug-in elements.

10.6.2 Electrical and Electronic Component Selection and Control Guidelines

Some of the guidelines are as follows [11]:

1. Find out the type of component needed to carry the specified function.

2. Determine the operational environment of the component.
3. Determine the availability and criticality of the component in question.
4. Determine the component's required reliability level and expected component stress when used in circuit.
5. Develop accurate procurement specification for the component in question.
6. For consistency with reliability prediction studies, make use of necessary derating factors.

10.7 SOFTWARE RELIABILITY

Today computers are playing a significant role in our day-to-day life. This role is following an increasing trend. In the United States Government fiscal year 1980, the amount spent on computer systems was about $57 billion [12]. Furthermore, over 50 percent of this amount was spent on computer software. This magnitude of money spent on software naturally leads one to believe that more attention should be given to software development so that it is reliable. Otherwise, the software errors may lead to catastrophic system failures. Therefore, this section briefly discusses two important aspects of software reliability. These are designing for reliability and mathematical modeling. Designing for reliability is basically concerned with structured programming. There are various definitions used to describe structured programming. However, a frequently used definition is that it is coding that avoids application of GO TO statements and program design that is TOP DOWN and modular [13]. Some of the advantages and disadvantages [12] of structured programming are described below.

Advantages

1. It is helpful in localizing an error.
2. Structured programming is especially clear.
3. It tends to restrict the coding to a handful of simple design procedures. Thus, it aids others in understanding the design.
4. It improves the productivity of the programmers.

Disadvantages

1. There may be reactionary attitudes of concerned personnel—in other words, reluctance of involved personnel to learn new approaches because they feel these procedures are not thoroughly proved.
2. Structured programs are inefficient in that, in some cases—in

comparison to their counterparts, the nonstructured programs—the structured programs need more running time or memory space.

Modular programming is practiced in a situation where a large programming problem is rather difficult to handle as an entity. In this case, the large program is divided into separate modules. The following are some of the advantages and disadvantages of modular programming [12]:

Advantages

1. It simplifies the debugging of the program.
2. It simplifies writing and management of the program.
3. It is less expensive and simpler to correct errors and change features after deployment.
4. It is quite compatible with the TOP-DOWN design.

Disadvantages

1. It demands more design effort.
2. It may need extra running time.
3. Many involved with programming are unwilling to try modular concept.
4. It may need extra memory.

Mathematical Modeling

This section presents one test-related reliability model [12]. The equation of the model is

$$Q_{fe} = Q_{ue} Q_{tm} \tag{10.41}$$

where Q_{fe} is the probability that an error occurs in the operational environments (i.e., field environment), Q_{ue} is the probability of the user exciting the error, and Q_{tm} is the probability of the tester missing the error. Thus, the probability Q_{ue} is given by

$$Q_{ue} = \frac{\alpha}{\beta} \tag{10.42}$$

where β is the number of module input values and α is the number of those module input values which cause errors.

Similarly, the probability Q_{tm} is given by

$$Q_{tm} = \left(1 - \frac{\alpha}{\beta}\right)^k \tag{10.43}$$

where k is the number of independent tests performed. Substituting Eq. (10.42)–(10.43) into Eq. (10.41) results in

$$Q_{fe} = \frac{\alpha}{\beta}\left(1 - \frac{\alpha}{\beta}\right)^k \quad (10.44)$$

The above equation can be used to calculate the approximate number of required tests when the reliability is given. The value of k must be much larger than the value of $1/Q_{ue}$ in order to make the value of Q_{fe} small.

10.8 SUMMARY

This chapter briefly presents various aspects of reliability used when designing engineering systems. The first topic discussed in the chapter is the failure data analysis. Under this topic, the maximum likelihood estimation technique is described to estimate parameters of component or system failure time probability distributions. The next topic of the chapter is reliability allocation. Here important benefits of the reliability allocation are outlined with two allocation approaches.

Reliability analysis is another topic covered in the chapter. Two commonly used reliability evaluation techniques at the equipment design stage are described. There are fault trees and failure modes and effects analysis.

A conceptually desirable approach, open to the designer, for achieving product reliability is described. In addition, the failure rate evaluation of electronic components and systems is briefly discussed.

Two other topics covered in the chapter are concerned with part selection and control and with ease of maintenance. Therefore, various useful guidelines associated with both these topics are listed.

The last topic of the chapter is software reliability. In this section, the advantages and disadvantages of both structured and modular programming are presented along with one test-related reliability model.

EXERCISES

1. The probability density function of failure times of an engineering system is given by

$$f(t) = \theta t e^{-(\theta t^2)/2} \quad (10.45)$$

where $f(t)$ is the failure probability density function, t is time, and θ is the parameter. Develop an expression for $\hat{\theta}$ with the aid of the maximum likelihood estimation technique. The symbol $\hat{\theta}$ signifies the maximum likelihood estimation of θ.

2. What are the advantages of reliability allocation?

3. An electronic system is composed of three independent units in series. The estimated constant failure rates for units 1, 2, and 3 are $\lambda_1 = 0.00004$, $\lambda_2 = 0.00005$, and $\lambda_3 = 0.00006$ failures/day, respectively. The entire series system specified constant failure rate is 0.00025 failures/hr. Compute the value of the failure rate to be allocated to each unit.
4. What are the areas you would look into at the equipment design stage which will help to simplify equipment maintenance in the field?
5. What is the main difference between failure modes and effects analysis (FMEA) and failure mode, effect and criticality analysis (FMECA)?
6. Compute the occurrence probability of the top event of the fault tree shown in Fig. 10.6. Occurrence probabilities of basic fault events are specified.

REFERENCES

1. D. K. Lloyd and M. Lipow, *Reliability: Management, Methods, and Mathematics,* Prentice-Hall, Englewood Cliffs, New Jersey, 1962, pp. 159–181.
2. J. R. King, *Probability Charts for Decision Making*, Industrial Press, New York, 1971.
3. P. D. T. O'Connor, *Practical Reliability Engineering*, Heyden, London, 1981, pp. 58–87.
4. M. L. Shooman, *Probabilistic Reliability: An Engineering Approach*, McGraw-Hill, New York, 1968.
5. Engineering Design Handbook: Development Guide for Reliability (Part II: Design for Reliability) (January 1976) AMCP 706-196, Headquarters US Army Material Command, Alexandria, Virginia 22333, pp. 5.1–5.33.
6. K. C. Kapur, and L. R. Lamberson, *Reliability in Engineering Design*, John Wiley & Sons, New York, 1977.
7. W. H. Von Alven, *Reliability Engineering*, Prentice-Hall, Englewood Cliffs, New Jersey, 1964, pp. 189–192.
8. J. S. Countinho, Failure-Effect Analysis, *Trans. N.Y. Acad. Sci.* Vol. 26(1964), (Series II), pp. 564–584.
9. H. J. H. Wassell, Reliability of Engineering Products, *Engineering Design Guides* 38, published for the Design Council, the British Standards Institution, and the Council of Engineering Institutions by Oxford University Press, Oxford, 1978, pp. 4–17.
10. US MIL-HBK-217, Reliability Prediction of Electronic Equipment, USAF Rome Air Development Center, available from the National Technical Information Service (NTIS), Springfield, Virginia, 1978.
11. RDH-376 Reliability Design Handbook, published by Reliability Analysis Center, USAF Rome Air Development Center, Griffiss Air Force Base, New York, 13441, 1976.

12. M. L. Shooman, *Software Engineering: Design, Reliability and Management*, McGraw-Hill, New York, 1983.
13. B. S. Dhillon and C. Singh, *Engineering Reliability: New Techniques and Applications*, John Wiley and Sons, New York, 1981, p. 126.

11

Time-Dependent Reliability Models

11.1 INTRODUCTION

The time-dependent mathematical models are widely used to predict reliability of a component or a system in day-to-day life. In the field of reliability there are various mathematical models which are developed to evaluate the reliability of a system or a component under different conditions. Therefore this chapter is devoted to reliability modeling of nonrepairable and repairable systems. Basically the chapter is divided into the following three parts:

1. Reliability evaluation of conventional networks
2. Reliability evaluation of three-state (two-failure-mode) device networks
3. Reliability evaluation of systems with common-cause failures

The first part is concerned with the reliability evaluation of commonly used conventional configurations in the reliability field such as series, parallel, k-out-of-m, and standby mathematical models. The components forming these networks are assumed to have only two states, that is, operational and failed states. In addition it is assumed that the failed system components are never repaired. The second part is devoted to the reliability evaluation of networks composed of devices with two failure modes. In other words, each device of the network may either be in normal operational state or failed in two mutually exclusive failure states I and II. The last portion of the chapter is concerned with the reliability prediction of systems with common-cause failures. In other words, the redundant system models take into consideration the occurrence of common-cause failures which may occur for various reasons, for example, operation, maintenance and design errors, external abnormal environments.

The above three topics are described in detail in the following sections.

11.2 RELIABILITY EVALUATION OF CONVENTIONAL NETWORKS

This section presents the time-dependent reliability analysis of commonly known reliability configurations composed of two-state devices. The following assumptions are common to all these configurations:

1. Each network element or component can be either in operating state or in failed state. In other words, each element is a two-state device.
2. Each network element fails independently.
3. All networks are composed of two-state devices.

11.2.1 K-out-of-m Units Network

This type of arrangement is frequently used to improve reliability of engineering systems especially in computer systems. The network is composed of m independent and active units. Therefore, this network simply represents a situation where at least k units out of m units must function normally for the system success. The well-known series and parallel networks in the reliability field are the special cases of this network at $k = m$ and $k = 1$, respectively. For the m identical units, the time-dependent reliability, $R_{k/m}(t)$, of the k-out-of-m units network is given by

$$R_{k/m}(t) = \sum_{j=k}^{m} \binom{m}{j} [1 - R(t)]^{m-j} [R(t)]^{j} \tag{11.1}$$

where

$$\binom{m}{j} \equiv \frac{m!}{j!(m-j)!} \tag{11.2}$$

and $R(t)$ is the network unit or element time, t, dependent reliability.

The reliability $R(t)$ of a single component [1] is given by

$$R(t) = \exp\left[-\int_0^t \lambda(x)\,dx\right] \tag{11.3}$$

where $\lambda(t)$ is known as the instantaneous failure rate or the hazard rate of a component.

Thus substituting Eq. (11.3) into Eq. (11.1) yields

$$R_{k/m}(t) = \sum_{j=k}^{m} \binom{m}{j} \left\{1 - \exp\left[-\int_0^t \lambda(x)\, dx\right]\right\}^{m-j} \times \exp\left[-j \int_0^t \lambda(x)\, dx\right] \tag{11.4}$$

This equation can be used to calculate reliability of the network whose components may follow increasing, decreasing, constant, and so on, hazard rates.

Special Case Model

In this model we assume that the failure rate of components in Eq. (11.4) is constant. In other words, the component failure times are exponentially distributed. Thus, we write

$$\lambda(t) = \lambda \tag{11.5}$$

where λ is the constant failure rate of a component.

Substituting the relationship (11.5) into Eq. (11.4) and integrating leads to

$$R_{k/m}(t) = \sum_{j=k}^{m} \binom{m}{j} (1 - e^{-\lambda t})^{m-j} e^{-j\lambda t} \tag{11.6}$$

The mean time to failure (MTTF) of a component or a system is given by

$$\text{MTTF} = \int_0^\infty R(t)\, dt \tag{11.7}$$

where $R(t)$ is the system or component time-dependent reliability.

Thus for constant failure rates of components, substituting Eq. (11.6) into Eq. (11.7) and integrating results in [1]

$$\text{MTTF}_{k/m} = \int_0^\infty R_{k/m}(t)\, dt = \sum_{j=k}^{m} (j\lambda)^{-1} \tag{11.8}$$

The reliability expressions for the set values of k and m in Eq. (11.6) are given in Table 11.1.

Table 11.1 Reliability Expressions for the k-out-of-m Unit Networks

Value of k	Value of m	k-out-of-m units system reliability
2	3	$R_{2/3}(t) = 3e^{-2\lambda t} - 2e^{-3\lambda t}$
3	4	$R_{3/4}(t) = 4e^{-3\lambda t} - 3e^{-4\lambda t}$
2	4	$R_{2/4}(t) = 3e^{-4\lambda t} - 8e^{-3\lambda t} + 6e^{-2\lambda t}$
$m-1$	m	$R_{(m-1)/m}(t) = me^{-(m-1)\lambda t} - (m-1)e^{-m\lambda t}$

The plots of Eq. (11.6) for $k = 3$, $m = 4$ and $k = 2$, $m = 3$ are shown in Fig. 11.1.

Example 11.1 An engineering system is composed of five active, identical, and independent units. At least three units must function normally for the system's success. The system starts operating at time $t = 0$ and each system

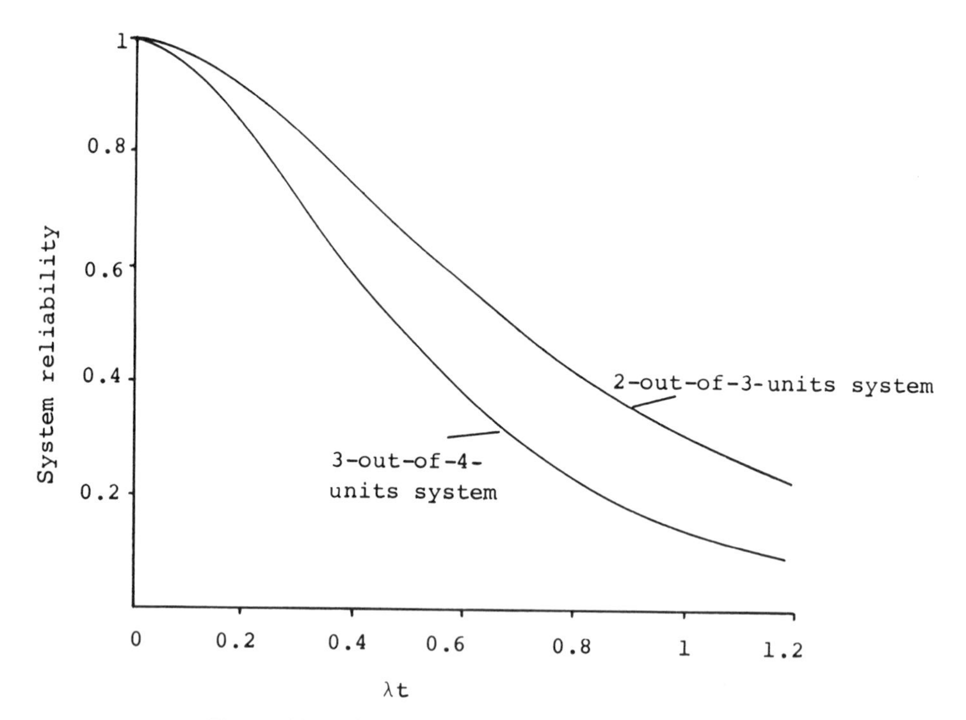

Figure 11.1 k-out-of-m unit system reliability plots.

unit failure rate λ is constant. Calculate the system reliability for a 200-hr mission if $\lambda = 0.007$ failures/hr.

Setting $k = 3$ and $m = 5$ in Eq. (11.6) leads to the 3-out-of-5 unit network reliability:

$$R_{3/5}(t) = 10e^{-3\lambda t} - 15e^{-4\lambda t} + 6e^{-5\lambda t} \tag{11.9}$$

Then using the given data, $\lambda = 0.007$ failures/hr and $t = 200$ hr in Eq. (11.9) results in

$$\begin{aligned} R_{3/5}(200) &= 10e^{-3(0.007)(200)} - 15e^{-4(0.007)(200)} \\ &\quad + 6e^{-5(0.007)(200)} \\ &= 0.15 - 0.056 + 0.0055 \\ &= 0.0995 \end{aligned}$$

Example 11.2 An electrical system consists of four active, identical, and independent units whose failure rates are constant. For the system's success at least three units must function normally. Each unit constant failure rate is 0.0005 failures/hr. Calculate the system mean time to failure.

In this example the data are defined as follows:

$$k = 3, \qquad m = 4, \qquad \text{and} \qquad \lambda = 0.0005 \text{ failures/hr}$$

Thus substituting the above data (i.e., $k = 3$ and $m = 4$) into Eq. (11.6) leads to

$$\begin{aligned} R_{3/4}(t) &= \sum_{j=3}^{4} \binom{4}{j} (1 - e^{-\lambda t})^{4-j} e^{-j\lambda t} \\ &= 4e^{-3\lambda t} - 3e^{-4\lambda t} \end{aligned} \tag{11.10}$$

Substituting Eq. (11.10) into Eq. (11.7) and integrating yields

$$\begin{aligned} \text{MTTF} &= \int_0^{\infty} (4e^{-3\lambda t} - 3e^{-4\lambda t})\, dt \\ &= \frac{7}{12\lambda} \end{aligned} \tag{11.11}$$

Substituting the value of $\lambda = 0.0005$ failures/hr into Eq. (11.11) results in

$$\text{MTTF} = \frac{7}{12(0.0005)} = 1166.67 \text{ hr}$$

11.2.2 Series Network

This network consists of m independent units in series. If any of the units fail, the series system fails. To predict worst case system reliability, it is assumed that all the system units form a series configuration. At $k = m$, in the k-out-of-m units system, the resulting network is the series network. Thus, from Eq. (11.1), the identical units series network reliability, $R_s(t)$, is given by

$$R_s(t) = R_{m/m}(t) = [R(t)]^m \tag{11.12}$$

For nonidentical m units, Eq. (11.12) is rewritten in the following form:

$$R_s(t) = \prod_{j=1}^{m} R_j(t) \tag{11.13}$$

where $R_j(t)$ is the jth component time-dependent reliability.

Substituting Eq. (11.3) for the jth unit into Eq. (11.13) yields

$$R_s(t) = \prod_{j=1}^{m} \exp\left[-\int_0^t \lambda_j(x)\,dx\right] \tag{11.14}$$

Special Case Model

For the jth unit constant failure rate, utilizing Eq. (11.5) in Eq. (11.14) and integrating leads to

$$R_s(t) = \prod_{j=1}^{m} e^{-\lambda_j t} \tag{11.15}$$

The series system mean time to failure is obtained by substituting Eq. (11.15) into Eq. (11.7) as follows:

$$\text{MTTF} = \int_0^\infty \left(\prod_{j=1}^{m} e^{-\lambda_j t}\right) dt$$

$$= \left(\sum_{j=1}^{m} \lambda_j\right)^{-1} \tag{11.16}$$

Example 11.3 An aircraft has three independent and identical engines. If any of the engines fails, the aircraft will crash. Each engine's constant failure rate is $\lambda = 0.0004$ failures/hr. If the aircraft starts flying at time $t = 0$, calculate its probability of flying successfully for a 15-hr continuous flying time. Assume that at the beginning of the mission the aircraft is as good as new.

Substituting the given data into Eq. (11.15) yields

$$R_s(15) = [e^{-(0.0004)(15)}]^3$$
$$= 0.9822$$

The reliability of the aircraft flying successfully is 0.9822.

11.2.3 Parallel Network

This is one of the commonly used configurations to improve system reliability. It is assumed that the network is composed of m active and independent units. For the system's success at least one unit must operate normally. At $k = 1$ in the k-out-of-m units network, the resulting configuration is the parallel network. Figure 11.2 shows the block diagram of the parallel network.

Setting $k = 1$ in Eq. (11.1) yields the identical unit parallel network reliability

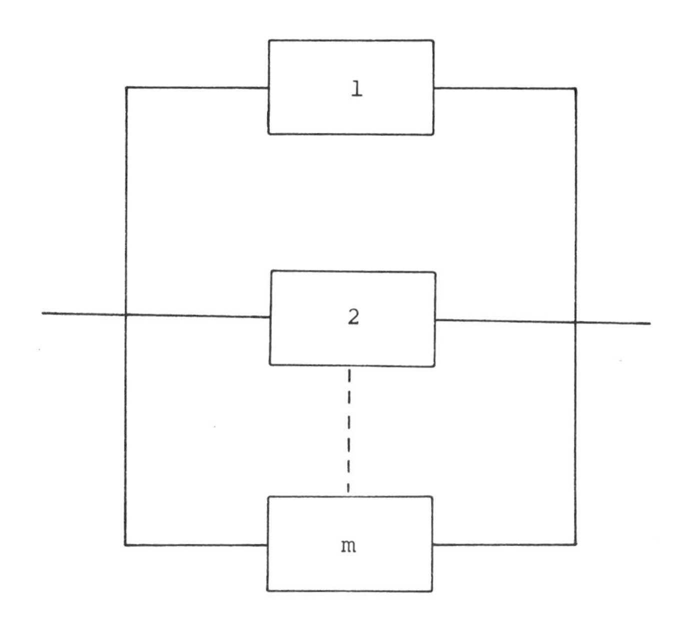

Figure 11.2 Parallel system block diagram.

$$R_p(t) = R_{1/m}(t) = \sum_{j=1}^{m} \binom{m}{j} [1 - R(t)]^{m-j} [R(t)]^j$$

$$= 1 - [1 - R(t)]^m \tag{11.17}$$

For nonidentical units, the above equation is rewritten in the following form:

$$R_p(t) = 1 - \prod_{j=1}^{m} F_j(t) \tag{11.18}$$

where

$$F_j(t) = [1 - R_j(t)] \tag{11.19}$$

and $R_j(t)$ is the jth unit or component time-dependent reliability.

By substituting Eq. (11.3) into Eq. (11.19), the resulting equation from Eq. (11.18) is as follows:

$$R_p(t) = 1 - \prod_{i=1}^{m} \left\{ 1 - \exp\left[- \int_0^t \lambda_i(x)\, dx \right] \right\} \tag{11.20}$$

Special Case Model

For the jth unit constant failure rate, utilizing Eq. (11.5) in Eq. (11.20) and integrating yields

$$R_p(t) = 1 - \prod_{i=1}^{m} (1 - e^{-\lambda_i t}) \tag{11.21}$$

For identical units, utilizing Eq. (11.21) in Eq. (11.7) and integrating leads to the following parallel network mean time to failure (MTTF_p) formula:

$$\mathrm{MTTF}_p = \int_0^\infty [1 - (1 - e^{-\lambda t})^m]\, dt$$

$$= \frac{1}{\lambda} \sum_{j=1}^{m} (j)^{-1} \tag{11.22}$$

For two identical units (i.e., $m = 2$), the plot of Eq. (11.21) is shown in Fig. 11.3.

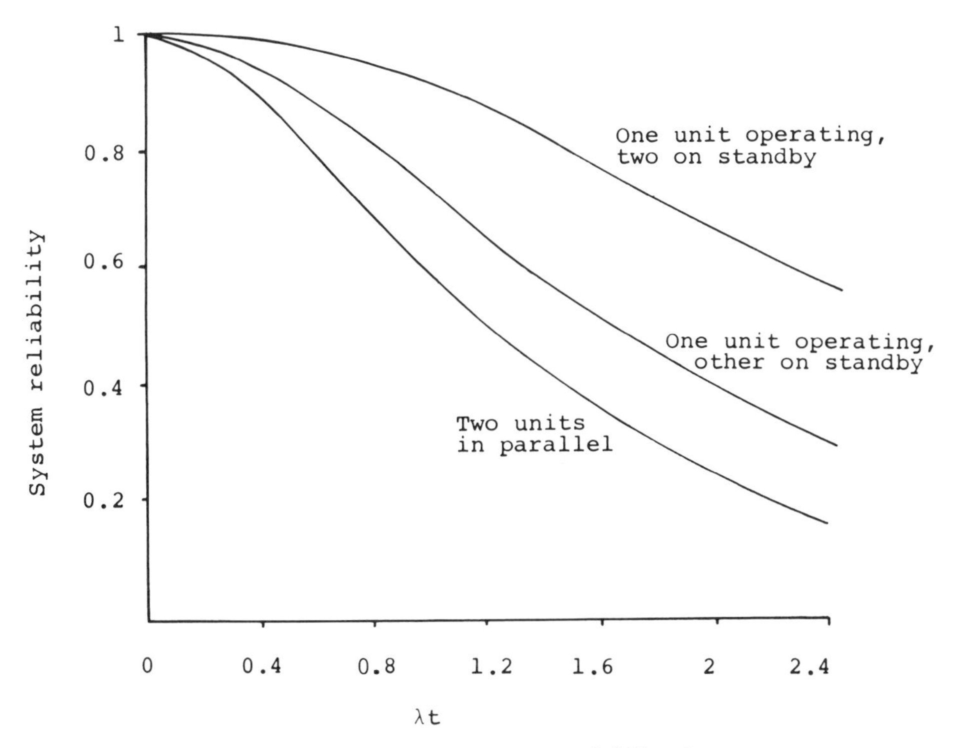

Figure 11.3 Redundant system reliability plots.

Example 11.4 An aircraft traffic control system consists of three active, independent, and identical central processing units. At least one central processing unit must function normally for the system's success. The constant failure rate λ of the central processing unit is 0.002 failures/hr. Calculate the air traffic control system mean time to failure.

By substituting the given data $m = 3$ and $\lambda = 0.002$ failures/hr in Eq. (11.22), the parallel system MTTF is

$$\begin{aligned}\text{MTTF} &= \frac{1}{(0.002)} \sum_{j=1}^{3} (j)^{-1} \\ &= \frac{1}{(0.002)} \left(1 + \frac{1}{2} + \frac{1}{3}\right) \\ &= 916.667 \text{ hr}\end{aligned}$$

11.2.4 Standby System

The standby system is another type of commonly used redundant configuration. In this situation the system is composed of g independent and identical units, out of which one is operating and the remaining $(g-1)$ are on the standby mission. As soon as the operational unit fails it is immediately replaced by one of the standbys. It is assumed that the switch to replace the failed unit, with one of the standbys, never fails and the standbys remain as good as new. The standby system time-dependent reliability $R_s(t)$ for unit constant failure rate λ from Ref. 2 is given by

$$R_s(t) = \sum_{j=0}^{g-1} \frac{(\lambda t)^j}{j!} e^{-\lambda t} \tag{11.23}$$

where t is time.

Substituting Eq. (11.23) into Eq. (11.7) and integrating leads to the following formula for the standby system mean time to failure (MTTF_s):

$$\text{MTTF}_s = \int_0^\infty \left[\sum_{j=0}^{(g-1)} \frac{(\lambda t)^j}{j!} e^{-\lambda t} \right] dt$$

$$= \frac{g}{\lambda} \tag{11.24}$$

Plots of Eq. (11.23) for $g = 2$ (one unit operating, other on standby) and for $g = 3$ (one unit operating, two on standby) are shown in Fig. 11.3.

Example 11.5 A system consists of three identical and independent electric motors. One motor is operating and the other two are on standby. Each motor's constant failure rate is $\lambda = 0.004$ failures/hr. Calculate the system reliability for a 50-hr mission by utilizing Eq. (11.23).

Thus setting $g = 3$, $\lambda = 0.004$ failures/hr, and $t = 50$ hr in Eq. (11.23) results in

$$R_s(50) = \sum_{j=0}^{2} \frac{[(0.004)(50)]^j}{j!} e^{-(0.004)(50)}$$

$$= e^{-(0.004)(50)} \left\{ 1 + (0.004)(50) + \frac{[(0.004)(50)]^2}{2} \right\}$$

$$= 0.8187(1 + 0.2 + 0.02)$$

$$= 0.9988$$

11.3 RELIABILITY EVALUATION OF SYSTEMS COMPOSED OF TWO-FAILURE-MODE (THREE-STATE) DEVICES

Many engineering systems or subsystems are composed of two-failure-mode devices. A two-failure-mode device is also known as a three-state device. Typical examples of a two-failure-mode device are an electronic diode, an electrical switch, a fluid flow valve, and so on. In the case of a diode it can fail either in a open or short mode whereas in the case of a fluid valve it may fail either in open or closed mode. To increase the reliability of systems composed of two-failure-mode devices, the concept of redundancy is used during the design phase of such systems. However, the increase in reliability of such redundant systems depends on the system configuration, the number of redundant devices, and the dominant failure mode of the device [3, 4].

Therefore, this section is concerned with the reliability evaluation of redundant configurations composed of two-failure-mode devices. In addition, reliability and availability analysis of a single two-failure-mode device are presented.

11.3.1 Reliability Evaluation of a Two-Failure-Mode Device

As mentioned earlier, a two-failure-mode device can be an electronic diode, an electrical switch, or a fluid flow valve, etc. Therefore this section is concerned with developing the reliability and failure mode probability equations for such a device by utilizing the Markov technique [2]. The device state-space diagram is shown in Fig. 11.4. The diagram depicts three states of the device, namely, normal operation state, failure mode I state, and failure mode II state. The failure mode I denotes the open mode failure state whereas the failure mode II denotes the short (closed) mode failure state.

The following assumptions are associated with the Markov model:

1. The transitional probability of an occurrence in a finite time interval from one state to another is generated by the product of λ_i times Δt, where λ_i is the ith mode constant failure rate (transition rate) of the device, for $i = 1, 2$, and Δt is the time interval.
2. The probabilities of occurrence of two or more transition occurrences during the time interval Δt are very small and can be ignored.
3. Failure rates are constant.
4. No repair is performed.

The following symbols are associated with the model:

i the ith state of the device; $i = 0$ (operating normally), $i = 1$ (failed in open mode), $i = 2$ (failed in short (closed) mode)

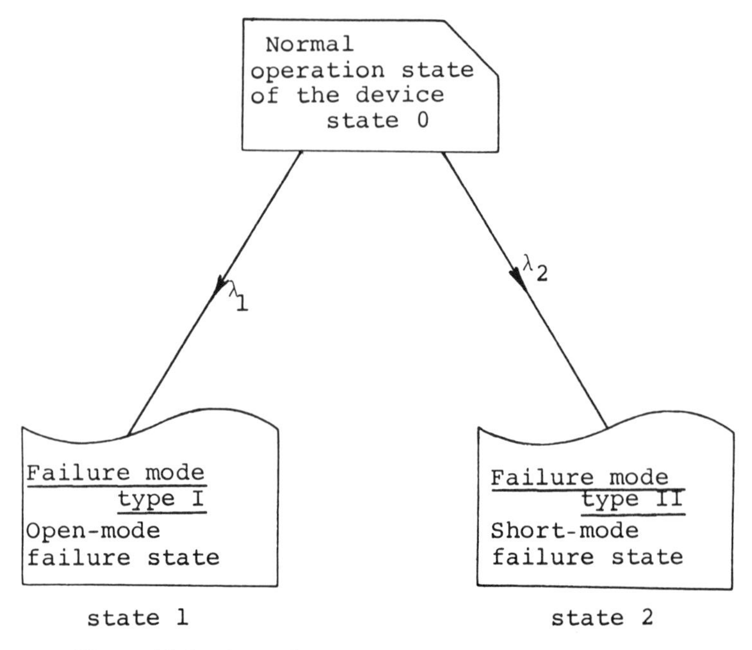

Figure 11.4 A two-failure-mode device state-space diagram.

$P_i(t)$ the probability that the system is in state i at time t, for $i = 0, 1, 2,$

λ_1 the open mode constant failure rate of the device (failure mode type I)

λ_2 the short (closed) mode constant failure rate of the device (failure mode type II)

Δt the time interval

s the Laplace transform variable

The following difference equations are associated with Fig. 11.4:

$$P_0(t + \Delta t) = P_0(t)(1 - \lambda_1 \Delta t)(1 - \lambda_2 \Delta t) \tag{11.25}$$

$$P_1(t + \Delta t) = P_1(t) + P_0(t)\lambda_1 \Delta t \tag{11.26}$$

$$P_2(t + \Delta t) = P_2(t) + P_0(t)\lambda_2 \Delta t \tag{11.27}$$

Rearranging Eqs. (11.25)–(11.27) yields

$$\frac{P_0(t + \Delta t) - P_0(t)}{\Delta t} = -P_0(t)\lambda_1 - P_0(t)\lambda_2 \tag{11.28}$$

$$\frac{P_1(t + \Delta t) - P_1(t)}{\Delta t} = P_0(t)\lambda_1 \tag{11.29}$$

$$\frac{P_2(t + \Delta t) - P_2(t)}{\Delta t} = P_0(t)\lambda_2 \tag{11.30}$$

Taking the limit of Eqs. (11.28)–(11.30) as $\Delta t \to 0$ results in

$$\frac{dP_0(t)}{dt} = -P_0(t)\lambda_1 - P_0(t)\lambda_2 \tag{11.31}$$

$$\frac{dP_1(t)}{dt} = P_0(t)\lambda_1 \tag{11.32}$$

$$\frac{dP_2(t)}{dt} = P_0(t)\lambda_2 \tag{11.33}$$

At $t = 0$, $P_0(0) = 1$ and $P_1(0) = P_2(0) = 0$

Taking the Laplace transforms of differential equations (11.31)–(11.33) leads to

$$sP_0(s) - P_0(0) = -P_0(s)\lambda_1 - P_0(s)\lambda_2 \tag{11.34}$$

$$sP_1(s) - P_1(0) = P_0(s)\lambda_1 \tag{11.35}$$

$$sP_2(s) - P_2(0) = P_0(s)\lambda_2 \tag{11.36}$$

After utilizing the given initial conditions and performing a few steps in between, Eqs. (11.34)–(11.36) become

$$P_0(s) = (s + \lambda_1 + \lambda_2)^{-1} \tag{11.37}$$

$$P_1(s) = \frac{\lambda_1(s + \lambda_1 + \lambda_2)^{-1}}{s} \tag{11.38}$$

$$P_2(s) = \frac{\lambda_2(s + \lambda_1 + \lambda_2)^{-1}}{s} \tag{11.39}$$

Taking the inverse Laplace transforms of Eqs. (11.37)–(11.39) leads to

$$P_0(t) = e^{-(\lambda_1+\lambda_2)t} \tag{11.40}$$

$$P_1(t) = \frac{\lambda_1}{\lambda_1 + \lambda_2}(1 - e^{-(\lambda_1+\lambda_2)t}) \tag{11.41}$$

$$P_2(t) = \frac{\lambda_2}{\lambda_1 + \lambda_2}(1 - e^{-(\lambda_1+\lambda_2)t}) \tag{11.42}$$

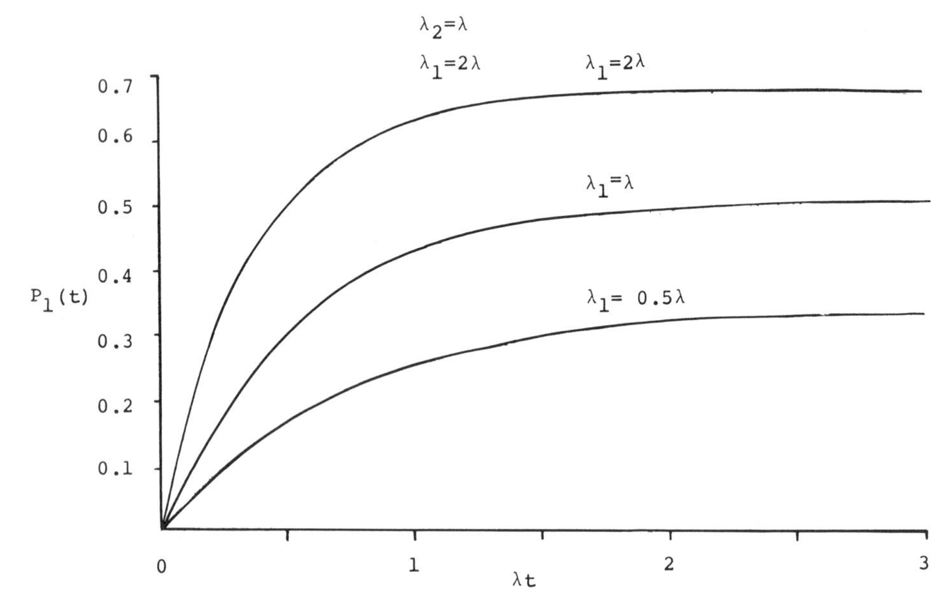

Figure 11.5 Failure mode type I probability plot.

The plot of Eq. (11.41) is shown in Fig. 11.5. The plot shows that as λt becomes large, the failure mode type I probability reaches a steady-state value. At $\lambda_1 = \lambda_2$, the steady-state value of $P_1(t)$ is equal to 0.5. In addition, as λ_1 becomes smaller the value of $P_1(t)$ decreases accordingly.

Example 11.6 An electronic system can fail in two mutually exclusive failure modes, i.e., type I (open mode) and type II (short mode). The open and short modes constant failure rates are $\lambda_1 = 0.002$ and $\lambda_2 = 0.004$ failures/hr, respectively. Calculate the value of the following items for a 100-hr mission:

1. System reliability
2. Open mode failure probability
3. Short mode failure probability

Utilizing the given data in Eq. (11.40) yields the system reliability $R(t)$ for a 100-hr mission:

$$R(100) = P_0(100) = e^{-(0.002+0.004)(100)}$$
$$= 0.5488$$

The system open mode failure probability from Eq. (11.41) is

$$P_1(100) = \frac{0.002}{0.002 + 0.004}(1 - e^{-(0.002+0.004)(100)})$$
$$= 0.1504$$

Similarly from Eq. (11.42) the system short mode failure probability is

$$P_2(100) = \frac{0.004}{0.002 + 0.004}(1 - e^{-(0.002+0.004)(100)})$$
$$= 0.3008$$

11.3.2 Reliability Evaluation of a Series System Composed of Two-Failure-Mode Devices

In an independent device series system, one element failing in an open mode will cause a complete system failure, whereas all elements of the system must malfunction in a short mode to induce failure for this system. The Markov technique is used to develop the reliability equation for the model.

Two-Unit Series System

For a two-independent-element series system, the Markov state space diagram is shown in Fig. 11.6.

The following assumptions are associated with the Markov model:

1. Device failures are statistically independent.
2. Device failure rates are constant.
3. Both devices are nonrepairable.

The following symbols are associated with the model:

$P_i(t)$ the probability that the system is in operating state i at time t, for $i = 0$ (both devices operating), $i = 1$ (device 1 failed in short mode, the other operating), $i = 2$ (device 2 failed in short mode, other operating)

λ_i the device "1" ith failure mode constant failure rate, for $i = 1$ (open mode); $i = 2$ (short mode)

λ_i' the device 2 ith failure mode constant failure rate, for $i = 1$ (open mode), $i = 2$ (short mode)

s the Laplace transform variable

To obtain the system reliability expression the differential equations associated with the model are

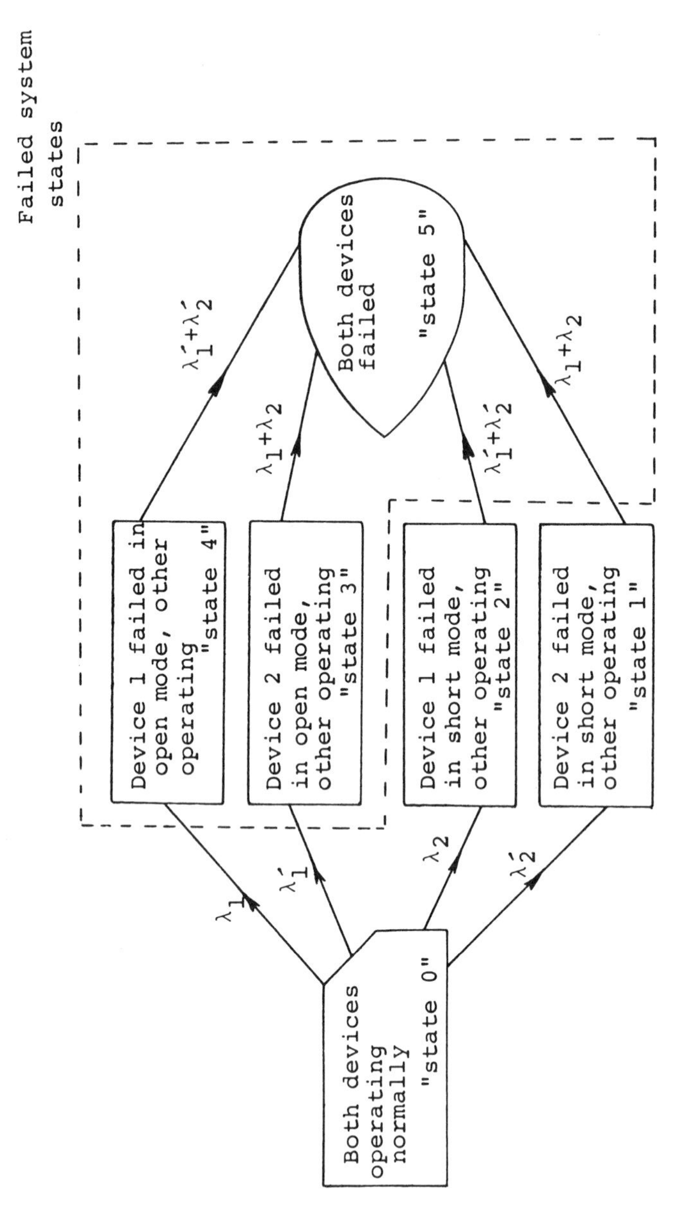

Figure 11.6 Series system state-space diagram.

$$\frac{dP_0(t)}{dt} + (\lambda_1 + \lambda_2 + \lambda_1' + \lambda_2')P_0 = 0 \tag{11.43}$$

$$\frac{dP_1(t)}{dt} + (\lambda_1 + \lambda_2)P_1(t) = P_0(t)\lambda_2' \tag{11.44}$$

$$\frac{dP_2(t)}{dt} + (\lambda_1' + \lambda_2')P_2(t) = P_0(t)\lambda_2 \tag{11.45}$$

At $t = 0$, $P_0(0) = 1$, other initial condition probabilities are equal to zero.

By utilizing the Laplace transform technique, the solutions of the above differential equations are as follows:

$$P_0(t) = e^{-(\lambda_1+\lambda_2+\lambda_1'+\lambda_2')t} \tag{11.46}$$

$$P_1(t) = \frac{\lambda_2'}{\lambda_1' + \lambda_2'}(e^{-(\lambda_1+\lambda_2)t} - e^{-(\lambda_1+\lambda_2+\lambda_1'+\lambda_2')t}) \tag{11.47}$$

$$P_2(t) = \frac{\lambda_2}{\lambda_1 + \lambda_2}(e^{-(\lambda_1'+\lambda_2')t} - e^{-(\lambda_1+\lambda_2+\lambda_1'+\lambda_2')t}) \tag{11.48}$$

For identical devices, that is $\lambda_1 = \lambda_1'$ and $\lambda_2 = \lambda_2'$, the above equations reduce to

$$P_0(t) = e^{-2(\lambda_1+\lambda_2)t} \tag{11.49}$$

$$P_1(t) = \frac{\lambda_2}{\lambda_1 + \lambda_2}(e^{-(\lambda_1+\lambda_2)t} - e^{-2(\lambda_1+\lambda_2)t}) \tag{11.50}$$

$$P_2(t) = \frac{\lambda_2}{\lambda_1 + \lambda_2}(e^{-(\lambda_1+\lambda_2)t} - e^{-2(\lambda_1+\lambda_2)t}) \tag{11.51}$$

Series system reliability $R(t)$ is given by

$$R(t) = P_0(t) + P_1(t) + P_2(t) \tag{11.52}$$

Thus substituting Eqs. (11.49)–(11.51) into Eq. (11.52) yields the two-identical-device series system reliability:

$$R(t) = e^{-2(\lambda_1+\lambda_2)t} + \frac{2\lambda_2}{\lambda_1 + \lambda_2}(e^{-(\lambda_1+\lambda_2)t} - e^{-2(\lambda_1+\lambda_2)t}) \tag{11.53}$$

The above equation may be written in the following form:

$$R(t) = \left[1 - \frac{\lambda_1}{\lambda_1 + \lambda_2}(1 - e^{-(\lambda_1+\lambda_2)t})\right]^2 - \left[\frac{\lambda_2}{\lambda_1 + \lambda_2}(1 - e^{-(\lambda_1+\lambda_2)t})\right]^2 \qquad (11.54)$$

Example 11.7 Two independent and identical electronic diodes form a series system. Each diode can fail in either open or short mode. Open and short modes constant failure rates are 0.005 failures/hr and 0.001 failures/hr, respectively. Calculate the series system reliability for a 200-hr mission.

By substituting the given data into Eq. (11.53), we get

$$\mathrm{R}(200) = e^{-2(0.005+0.001)(200)} + \frac{2(0.001)}{(0.005 + 0.001)}(e^{-(0.005+0.001)(200)} - e^{-2(0.005+0.001)(200)})$$

$$= 0.0907 + 0.3333(0.3012 - 0.0907)$$

$$= 0.161$$

To check the correctness of Eq. (11.54) we substitute the given data in that equation as follows:

$$R(200) = \left[1 - \frac{0.005}{0.005 + 0.001}(1 - e^{-(0.005+0.001)(200)})\right]^2 - \left[\frac{0.001}{0.005 + 0.001}(1 - e^{-(0.005+0.001)(200)})\right]^2$$

$$= 0.1744 - 0.0136$$

$$= 0.161$$

Therefore, it may be concluded that utilizing either Eq. (11.53) or Eq. (11.54) yields the same end result.

Three-Unit Series System

Similarly as for the two-unit series system, one can develop a reliability expression for a three independent and identical unit system. For device constant failure rates and identical devices, the system reliability is given by

$$R(t) = \left[1 - \frac{\lambda_1}{\lambda_1 + \lambda_2}(1 - e^{-(\lambda_1+\lambda_2)t})\right]^3 - \left[\frac{\lambda_2}{\lambda_1 + \lambda_2}(1 - e^{-(\lambda_1+\lambda_2)t})\right]^3 \quad (11.55)$$

In Eq. (11.55) a device open and short mode failure probabilities, respectively, are given by

$$q_0(t) = \frac{\lambda_1}{\lambda_1 + \lambda_2}(1 - e^{-(\lambda_1+\lambda_2)t}) \quad (11.56)$$

$$q_s(t) = \frac{\lambda_2}{\lambda_1 + \lambda_2}(1 - e^{-(\lambda_1+\lambda_2)t}) \quad (11.57)$$

One should note here that Eqs. (11.56)–(11.57) are the same as Eqs. (11.41)–(11.42), respectively.

n *Unit Series System*

For n identical devices, the time-dependent series system reliability from Eq. (11.55) is generalized to the following form:

$$R(t) = \left[1 - \frac{\lambda_1}{\lambda_1 + \lambda_2}(1 - e^{-(\lambda_1+\lambda_2)t})\right]^n - \left[\frac{\lambda_2}{\lambda_1 + \lambda_2}(1 - e^{-(\lambda_1+\lambda_2)t})\right]^n \quad (11.58)$$

After substituting Eqs. (11.56)–(11.57) into Eq. (11.58), the resulting equation is rewritten in the following form:

$$R(t) = [1 - q_0(t)]^n - [q_s(t)]^n \quad (11.59)$$

For nonidentical devices, the series system reliability equation (11.59) becomes

$$R(t) = \prod_{i=1}^{n} [1 - q_{0i}(t)] - \prod_{i=1}^{n} q_{si}(t) \quad (11.60)$$

where $q_{0i}(t)$ is the ith device open mode failure probability at time t and $q_{si}(t)$ is the ith device short mode failure probability at time t.

11.3.3 Reliability Evaluation of a Parallel System Composed of Two-Failure-Mode Devices

This configuration is the dual of the series configuration. For a parallel system to fail, all the elements must fail in the open mode or one of the elements must fail in a short mode to cause the system to fail completely. Because of duality for independent and identical devices with constant failure rates, the parallel system reliability, $R(t)$, from Eq. (11.58) is

$$R(t) = \left[1 - \frac{\lambda_2}{\lambda_1 + \lambda_2}(1 - e^{-(\lambda_1+\lambda_2)t})\right]^m - \left[\frac{\lambda_1}{\lambda_1 + \lambda_2}(1 - e^{-(\lambda_1+\lambda_2)t})\right]^m \tag{11.61}$$

where m is the number of devices connected in parallel.

With the aid of Eqs. (11.56)–(11.57) the above equation is rewritten in the following form:

$$R(t) = [1 - q_s(t)]^m - [q_0(t)]^m \tag{11.62}$$

For nonidentical devices, the above equation becomes

$$R(t) = \prod_{i=1}^{m} [1 - q_{si}(t)] - \prod_{i=1}^{m} q_{0i}(t) \tag{11.63}$$

Example 11.8 Suppose the two independent electronic diodes of Example 11.7 form an independent unit parallel system instead of a series system. Compute the system reliability for the given data in Example 11.7.

A two-unit parallel system reliability from Eq. (11.61) is

$$R(t) = \left[1 - \frac{\lambda_2}{\lambda_1 + \lambda_2}(1 - e^{-(\lambda_1+\lambda_2)t})\right]^2 - \left[\frac{\lambda_1}{\lambda_1 + \lambda_2}(1 - e^{-(\lambda_1+\lambda_2)t})\right]^2 \tag{11.64}$$

Substituting the given data in the above equation results in

$$R(100) = \left[1 - \frac{0.001}{0.005 + 0.001}(1 - e^{-(0.005+0.001)(200)})\right]^2 - \left[\frac{0.005}{0.005 + 0.001}(1 - e^{-(0.005+0.001)(200)})\right]^2$$

$$= (0.8835)^2 - (0.5823)^2$$
$$= 0.7806 - 0.3391 = 0.4415$$

The parallel system reliability for a 200-hr mission is 0.4415.

11.3.4 Reliability Analysis of a Repairable Two-Failure-Mode Device

Generally whenever a device fails, it is repaired back to its fully operational state. Therefore, in this section we extent analysis to take into consideration the repair of a two-failure-mode device. A nonrepairable device state-space diagram is shown in Fig. 11.4. From now on whenever the device fails in either failure mode I or failure mode II, it is repaired back to its successfully operating state. For example in Fig. 11.4, the device is repaired at a constant rate, β_1, from failure mode I and at a constant rate, β_2, from failure mode II. In addition to assumptions (1)–(3) already outlined for the nonrepairable device in the chapter, the following additional assumptions are associated with the repairable two-failure-mode device model:

4. The device is repairable.
5. The repaired device is as good as new.
6. The device repair rates are constant.

In addition to symbols already defined for the nonrepairable device (see Fig. 11.4), the following new symbols are introduced:

β_1 the constant repair rate from failure mode type I
β_2 the constant repair rate from failure mode type II

The following equations are associated with the repairable device Markov model:

$$\frac{dP_0(t)}{dt} + (\lambda_1 + \lambda_2)P_0(t) - P_1(t)\beta_1 - P_2(t)\beta_2 = 0 \qquad (11.65)$$

$$\frac{dP_1(t)}{dt} + \beta_1 P_1(t) - P_0(t)\lambda_1 = 0 \qquad (11.66)$$

$$\frac{dP_2(t)}{dt} + \beta_2 P_2(t) - P_0(t)\lambda_2 = 0 \qquad (11.67)$$

$$\text{At } t = 0, \qquad P_0(0) = 1, \qquad P_1(0) = P_2(0) = 0$$

The Laplace transforms of the solution [3] to differential equations (11.65)–(11.67) are as follows:

$$P_0(s) = \frac{(s+\beta_1)(s+\beta_2)}{s[s^2 + s(\beta_1+\beta_2+\lambda_1+\lambda_2) + (\beta_1\beta_2+\lambda_1\beta_2+\lambda_2\beta_1)]} \tag{11.68}$$

$$P_1(s) = \frac{\lambda_1(s+\beta_1)}{s[s^2 + s(\beta_1+\beta_2+\lambda_1+\lambda_2) + \beta_1\beta_2+\lambda_1\beta_2+\lambda_2\beta_1]} \tag{11.69}$$

$$P_2(s) = \frac{\lambda_2(s+\beta_1)}{s[s^2 + s(\beta_1+\beta_2+\lambda_1+\lambda_2) + \beta_1\beta_2+\lambda_1\beta_2+\lambda_2\beta_1]} \tag{11.70}$$

To find the steady-state probabilities, the following final value theorem can be utilized:

$$\lim_{t\to\infty} f(t) = \lim_{s\to 0} sf(s) \tag{11.71}$$

Thus by utilizing Eqs. (11.68)–(11.70) in the right-hand side of the theorem, (11.71), the following steady-state probability equations result:

$$P_0 = \beta_1\beta_2/(\beta_1\beta_2+\lambda_1\beta_2+\lambda_2\beta_1) \tag{11.72}$$

$$P_1 = \lambda_1\beta_2/(\beta_1\beta_2+\lambda_1\beta_2+\lambda_2\beta_1) \tag{11.73}$$

$$P_2 = \lambda_2\beta_1/(\beta_1\beta_2+\lambda_1\beta_2+\lambda_2\beta_1) \tag{11.74}$$

The above equations can be used to calculate steady-state probabilities of a repairable two-failure-mode device. The steady-state availability of the device is given by Eq. (11.72).

Example 11.9 An engineering system can only fail in two mutually exclusive failure modes I and II. The constant failure rates of modes I and II are $\lambda_1 = 0.003$ and $\lambda_2 = 0.005$ failures/hr, respectively. Similarly, the corresponding constant repair rates from failure modes I and II are $\beta_1 = 0.004$ and $\beta_2 = 0.006$ repairs/hr. Calculate the system steady-state availability.

Substituting the above specified data into Eq. (11.72) yields the following system steady-state availability:

$$A = P_0 = \frac{(0.004)(0.006)}{(0.004)(0.006) + (0.003)(0.006) + (0.005)(0.004)}$$

$$= 0.3871$$

The two-failure-mode system steady-state availability is 0.3871.

11.4 RELIABILITY EVALUATION OF SYSTEMS WITH COMMON-CAUSE FAILURES

In the last decade attention has been given to the subject of common-cause failures. The main reason for this attention was that the multiple units of many engineering systems failed due to a single cause. In other words, the common-cause failures have received widespread attention in reliability analysis of redundant components because the assumption of statistical independent failure of redundant units is easily violated in practice [5–7]. Reference 6 reports on the frequency of common-cause failures in the U.S. nuclear power reactor industry: "of 379 component failures or groups of failures arising from independent causes, 78 involved common-cause failures of two or more components." The following are generally the main reasons of common-cause failures [5]:

1. Fire, flood, earthquake, and tornado
2. Temperature, dust, dirt, and humidity
3. Operation and maintenance errors
4. Design and functional deficiency

The published literature on common-cause failures is listed in Ref. 8. This section presents various mathematical models which take into consideration the occurrence of common-cause failures.

11.4.1 Reliability Evaluation of a Redundant System with Common-Cause Failures

Basically this Markov model is concerned with the reliability evaluation of a m-out-of-g unit system. More clearly the system is composed of g identical units out of which at least m units are required to function successfully for the system's success. In addition, the system may fail due to the occurrence of common-cause failures. The state-space diagram of the system is given in Fig. 11.7.

The following assumptions are associated with the Markov model:

1. System common-cause and other failure rates are constant.
2. Common-cause and other failures are statistically independent.
3. Common-cause failures can only occur with more than one unit.
4. System units are never repaired.
5. All units are identical and active.

The following symbols are associated with the model:

λ_i constant failure rate of the system when the system is in state i, $i = 0, 1, 2, 3, \ldots, (n-1)$

λ constant failure rate of a unit

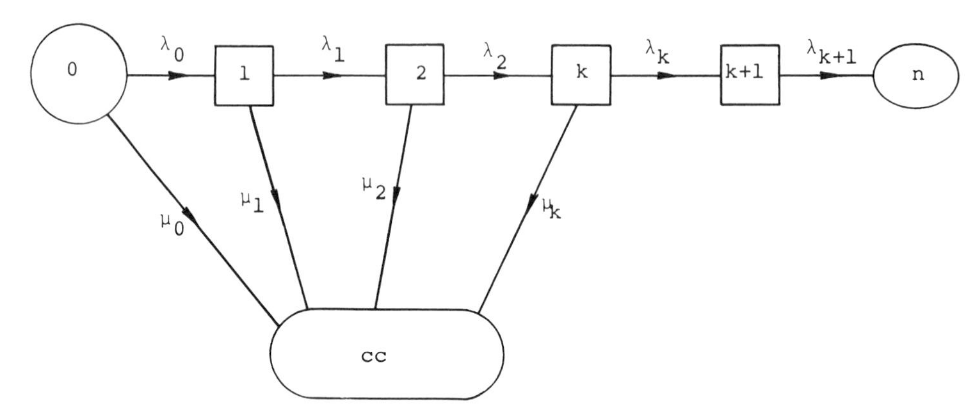

Figure 11.7 System transition diagram.

μ_i — constant common-cause failure rate of the system when in state i, $i=0, 1, 2, \ldots, k$

i — state of the unfailed system (for 1-out-of-g unit system), number of failed units, $i=0, 1, 2, 3, \ldots, (n-1)$

j — state of the failed system (for 1-out-of-g unit system), $j=n$ means failure not due to a common cause; $j=cc$ means failure due to common cause

$P_i(t)$ — probability that system is in unfailed state i at time t

$P_n(t)$ — probability that system is in failed state n (not due to a common cause) at time t

$P_{cc}(t)$ — probability that system is in failed state at time t due to a common-cause failure

s — Laplace transform variable

N — the number of identical units

The differential equations for the model shown in Fig. 11.7 are

$$\frac{dP_0(t)}{dt} = -(\lambda_0 + \mu_0)P_0(t) \tag{11.75}$$

$$\frac{dP_1(t)}{dt} = -(\lambda_1 + \mu_1)P_1(t) + P_0(t)\lambda_0 \tag{11.76}$$

$$\frac{dP_2(t)}{dt} = -(\lambda_2 + \mu_2)P_2(t) + P_1(t)\lambda_1 \tag{11.77}$$

$$\vdots \qquad\qquad\qquad \vdots$$

$$\frac{dP_k(t)}{dt} = -(\lambda_k + \mu_k)P_k(t) + P_{k-1}(t)\lambda_{k-1}$$

$$\text{for } k = 1, 2, 3, \ldots, n-2 \qquad (11.78)$$

$$\frac{dP_{k+1}(t)}{dt} = -\lambda_{(n-1)}P_{k+1}(t) + P_k(t)\lambda_k \qquad \text{for } k = n-2 \qquad (11.79)$$

$$\vdots$$

$$\frac{dP_n(t)}{dt} = \lambda_{(n-1)}P_{k+1}(t) \qquad (11.80)$$

$$\frac{dP_{cc}(t)}{dt} = \sum_{i=0}^{k} \mu_i P_i(t) \qquad \text{for } n = N, \quad k = n-2, \quad N \geq 2 \qquad (11.81)$$

$$\lambda_i = (n-i)\lambda \qquad \text{for } i = 0, 1, 2, \ldots, n-1$$

At $t = 0$, $P_0(0) = 1$ other initial condition probabilities are equal to zero.

The Laplace transforms of the solution are Eqs. (11.82)–(11.86):

$$P_0(s) = \frac{1}{(s + \lambda_0 + \mu_0)} \qquad (11.82)$$

$$P_k(s) = \frac{\prod_{i=0}^{k-1} \lambda_i}{\prod_{i=0}^{k} (s + \lambda_i + \mu_i)} \qquad \text{for } k = 1, 2, 3, \ldots, n-2 \qquad (11.83)$$

$$P_{k+1}(s) = \frac{\prod_{i=0}^{k} \lambda_i}{\prod_{i=0}^{k} (s + \lambda_i + \mu_i)(s + \lambda_{k+1})} \qquad \text{for } k = n-2 \qquad (11.84)$$

$$P_n(s) = \frac{\prod_{i=0}^{k+1} \lambda_i}{\prod_{i=0}^{k} (s + \lambda_i + \mu_i)(s + \lambda_{k+1})s} \qquad (11.85)$$

$$P_{cc}(s) = \left(\frac{1}{s}\right)\sum_{i=0}^{k} P_i(s)\mu_i \qquad (11.86)$$

Special Case Model

This model is the special case of the above model. The model represents a two active identical and independent unit system which may fail due to the occurrence of common-cause failures [9]. Then by setting $N=2$, Eqs. (11.82)–(11.86) yield

$$P_0(s) = (s + \lambda_0 + \mu_0)^{-1} \tag{11.87}$$

where

$$\lambda_0 \equiv 2\lambda$$

$$P_1(s) = \frac{\lambda_0}{(s + \lambda_1)(s + \lambda_0 + \mu_0)} \tag{11.88}$$

where

$$\lambda_1 \equiv \lambda$$

$$P_2(s) = \frac{\lambda_0 \lambda_1}{(s + \lambda_1)(s + \lambda_0 + \mu_0)s} \tag{11.89}$$

$$P_{cc}(s) = \frac{\mu_0}{s(s + \lambda_0 + \mu_0)} \tag{11.90}$$

Taking the inverse Laplace transforms of Eqs. (11.87)–(11.90) results in

$$P_0(t) = e^{-(\lambda_0 + \mu_0)t} \tag{11.91}$$

$$P_1(t) = \left(\frac{\lambda_0}{\lambda_0 + \mu_0 - \lambda_1}\right)(e^{-\lambda_1 t} - e^{-(\lambda_0 + \mu_0)t}) \tag{11.92}$$

$$P_2(t) = \left(\frac{\lambda_0}{\lambda_0 + \mu_0}\right) + \left(\frac{\lambda_0}{\lambda_0 + \mu_0 - \lambda_1}\right)\left(\frac{\lambda_1}{\lambda_0 + \mu_0} e^{-(\lambda_0 + \mu_0)t} - e^{-\lambda_1 t}\right) \tag{11.93}$$

$$P_{cc}(t) = \left(\frac{\mu_0}{\lambda_0 + \mu_0}\right)(1 - e^{-(\lambda_0 + \mu_0)t}) \tag{11.94}$$

The reliability, $R(t)$, is obtained by summing Eqs. (11.91)–(11.92):

$$R(t) = \frac{\lambda_0}{(\lambda_0 + \mu_0 - \lambda_1)} e^{-\lambda_1 t} + \frac{(\mu_0 - \lambda_1)}{(\lambda_0 + \mu_0 - \lambda_1)} e^{-(\lambda_0 + \mu_0)t} \tag{11.95}$$

Since $\lambda_0 = 2\lambda$ and $\lambda_1 = \lambda$, the above equation is rewritten in the following form:

$$R(t) = \frac{2\lambda}{2\lambda + \mu_0 - \lambda} e^{-\lambda t} + \frac{(\mu_0 - \lambda)}{2\lambda + \mu_0 - \lambda} e^{-(2\lambda+\mu_0)t} \quad (11.96)$$

For the specified values of λ and μ, the time-dependent reliability plot of Eq. (11.96) is shown in Fig. 11.8. The plot shows that as t increases the system reliability decreases accordingly. Similarly for the increasing value of the constant common-cause failure rate, the system reliability also decreases.

The system mean time to failure (MTTF) is obtained by substituting Eq. (11.96) into Eq. (11.7) as follows:

$$\begin{aligned} \text{MTTF} &= \int_0^\infty \left[\frac{2\lambda}{(2\lambda + \mu_0 - \lambda)} e^{-\lambda t} + \frac{(\mu_0 - \lambda)}{(2\lambda + \mu_0 - \lambda)} e^{-(2\lambda+\mu_0)t} \right] dt \\ &= \frac{2\lambda}{2\lambda + \mu_0 - \lambda} \int_0^\infty e^{-\lambda t}\, dt + \frac{(\mu_0 - \lambda)}{2\lambda + \mu_0 - \lambda} \int_0^\infty e^{-(2\lambda+\mu_0)t}\, dt \\ &= \frac{2}{2\lambda + \mu_0 - \lambda} + \frac{(\mu_0 - \lambda)}{(2\lambda + \mu_0 - \lambda)(2\lambda + \mu_0)} \\ &= \frac{3}{2\lambda + \mu_0} \end{aligned} \quad (11.97)$$

Example 11.10 A system contains two active, identical, and independent units. The system may fail due to the occurrence of common-cause failures. In addition, the common-cause failures may only occur when more than one unit is in the system. The system common-cause failure rate is 0.002 failures/hr and the unit failure rate is 0.02 failures/hr. Calculate the system mean time to failure. By utilizing the given data in Eq. (11.97) we get

$$\begin{aligned} \text{MTTF} &= \frac{3}{2(0.02) + 0.002} \\ &= 71.43 \text{ hr} \end{aligned}$$

11.4.2 Reliability Analysis of a Repairable Two-Unit Redundant System

This Markov model represents a repairable redundant system with common-cause failures. The system consists of two active, independent, and identical units which may fail due to the occurrence of common-cause failures. The system is only repaired when at least one unit is functioning normally. The

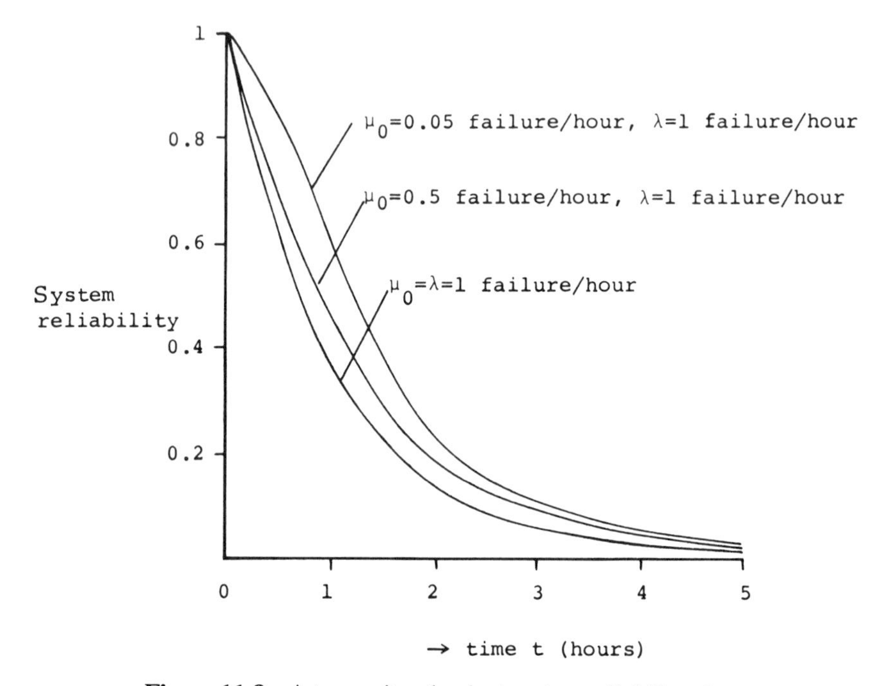

Figure 11.8 A two-unit redundant system reliability plot.

failed system is never repaired. The system transition diagram is shown in Fig. 11.9. The following assumptions are associated with the Markov model:

1. Only one unit is required for the system's success.
2. Common-cause and other failures are statistically independent.
3. Common-cause failures can only occur with more than one unit.
4. Common-cause and other failure rates are constant.
5. The repaired unit is as good as new.
6. Failed unit repair rate is constant.
7. Both units are identical; system fails when both units fail.

The following symbols are used to develop differential equations for the Markov model:

$P_i(t)$	probability that the system is in state i at time t, $i = 0, 1, 2, cc$
λ	unit constant failure rate
μ	unit constant repair rate
α	system common-cause failure rate
s	the Laplace transform variable

The following system of differential equations are associated with Fig. 11.9:

$$\frac{dP_0(t)}{dt} + (2\lambda + \alpha)P_0(t) = P_1(t)\mu \tag{11.98}$$

$$\frac{dP_1(t)}{dt} + (\mu + \lambda)P_1(t) = P_0(t)2\lambda \tag{11.99}$$

$$\frac{dP_2(t)}{dt} = P_1(t)\lambda \tag{11.100}$$

$$\frac{dP_{cc}(t)}{dt} = \alpha P_0(t) \tag{11.101}$$

At $t = 0$, $P_0(0) = 1$, $P_1(0) = P_2(0) = P_{cc}(0) = 0$

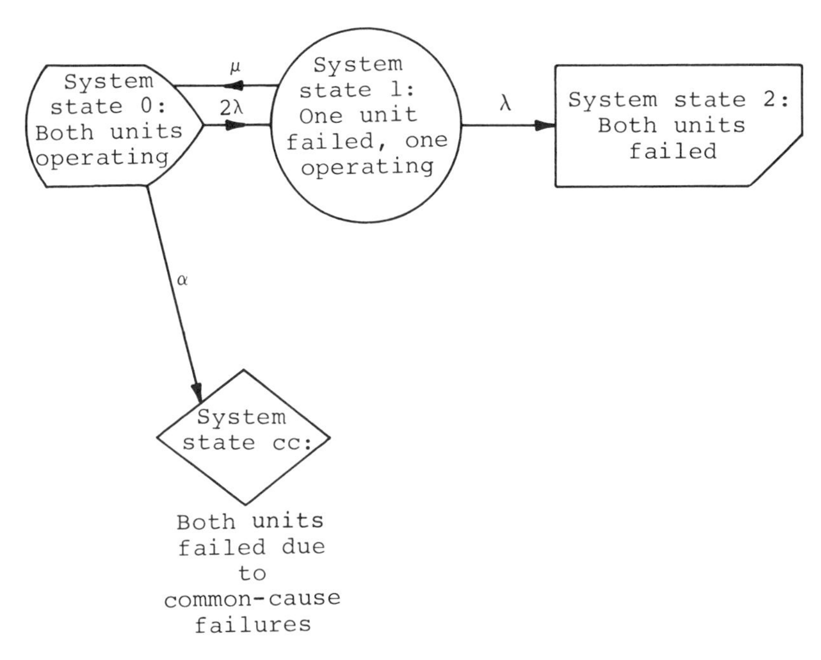

Figure 11.9 System state-space diagram.

The Laplace transforms of the solution are Eqs. (11.102)–(11.105):

$$P_0(s) = \frac{s + \lambda + \mu}{(s + 2\lambda + \alpha)(s + \lambda + \mu) - 2\lambda\mu} \tag{11.102}$$

$$P_1(s) = 2\lambda/[(s + 2\lambda + \alpha)(s + \lambda + \mu) - 2\lambda\mu] \tag{11.103}$$

$$P_2(s) = 2\lambda^2/s[(s + 2\lambda + \alpha)(s + \lambda + \mu) - 2\lambda\mu] \tag{11.104}$$

$$P_{cc}(s) = \alpha(s + \lambda + \mu)/s[(s + 2\lambda + \alpha)(s + \lambda + \mu) - 2\lambda\mu] \tag{11.105}$$

The above equations can easily be inverted into the time domain. The system mean time to failures (MTTF) can be obtained from the following theorem:

$$\text{MTTF} = \lim_{s \to 0} R(s) \tag{11.106}$$

where $R(s)$ denotes the Laplace transform of system reliability.

Thus, by summing Eqs. (11.102) and (11.103) the Laplace transform of the system reliability is

$$R(s) = P_0(s) + P_1(s)$$
$$= \frac{s + \lambda + \mu + 2\lambda}{(s + 2\lambda + \alpha)(s + \lambda + \mu) - 2\lambda\mu} \tag{11.107}$$

Thus, substituting Eq. (11.107) into Eq. (11.106) and letting $s \to 0$ yields

$$\text{MTTF} = \frac{3\lambda + \mu}{(2\lambda + \alpha)(\lambda + \mu) - 2\lambda\mu} \tag{11.108}$$

11.4.3 Availability Analysis of a Two-Nonidentical-Unit System with Common-Cause Failures

This model represents an active two-nonidentical-unit system [10], which may fail due to common-cause failures. Whenever a unit fails it is repaired. In addition, only one unit is needed for the system's success.

The system space diagram is shown in Fig. 11.10. The following assumptions are associated with the Markov model:

1. Unit failure and repair rates are constant.
2. System common-cause failure rate is constant.
3. Common-cause and other failures are statistically independent.

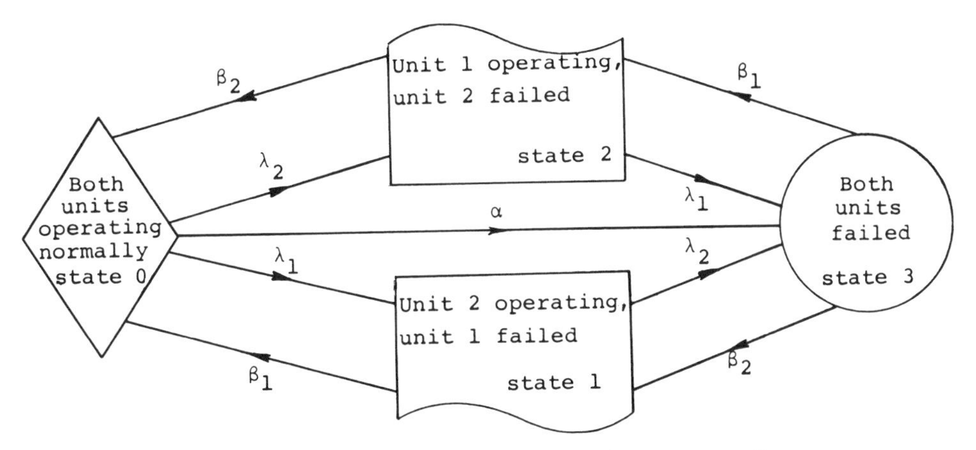

Figure 11.10 Two nonidentical redundant unit state-space diagram.

4. Common-cause failures can only occur with more than one unit.
5. The repaired unit is as good as new.
6. Both units are active.

The following symbols are associated with the model:

$P_i(t)$ is the probability that the system is in state i at time t, $i=0$ (both units operating), $i=1$ (unit 2 operating, unit 1 failed), $i=2$ (unit 1 operating, unit 2 failed), $i=3$ (both units failed).

λ_i is the ith unit constant failure rate; $i=1$ (unit 1), $i=2$ (unit 2).

β_i is the ith unit constant repair rate; $i=1$ (unit 1) $i=2$ (unit 2).

α is the system constant common-cause failure rate.

The following equations are associated with Fig. 11.10:

$$\frac{dP_0(t)}{dt} + (\lambda_1 + \lambda_2 + \alpha)P_0(t) = P_1(t)\beta_1 + P_2(t)\beta_2 \tag{11.109}$$

$$\frac{dP_1(t)}{dt} + (\lambda_2 + \beta_1)P_1(t) = P_3(t)\beta_2 + P_0(t)\lambda_1 \tag{11.110}$$

$$\frac{dP_2(t)}{dt} + (\lambda_1 + \beta_2)P_2(t) = P_0(t)\lambda_2 + P_3(t)\beta_1 \tag{11.111}$$

$$\frac{dP_3(t)}{dt} + (\beta_1 + \beta_2)P_3(t) = P_0(t)\alpha + P_1(t)\lambda_2 + P_2(t)\lambda_1 \qquad (11.112)$$

$$\text{at } t = 0, \qquad P_0(0) = 1, \qquad P_1(0) = P_2(0) = P_3(0) = 0$$

Setting the derivatives in Eqs. (11.109)–(11.112) equal to zero results in

$$(\lambda_1 + \lambda_2 + \alpha)P_0 - P_1\beta_1 - P_2\beta_2 = 0 \qquad (11.113)$$

$$(\lambda_2 + \beta_1)P_1 - P_3\beta_2 - P_0\lambda_1 = 0 \qquad (11.114)$$

$$(\lambda_1 + \beta_2)P_2 - P_0\lambda_2 - P_3\beta_1 = 0 \qquad (11.115)$$

$$(\beta_1 + \beta_2)P_3 - P_0\alpha - P_1\lambda_2 - P_2\lambda_1 = 0 \qquad (11.116)$$

$$P_0 + P_1 + P_2 + P_3 = 1 \qquad (11.117)$$

Solving Eqs. (11.113)–(11.117) yields the following steady-state probability equations:

$$P_0 = \frac{B_1\beta_1\beta_2}{B_2} \qquad (11.118)$$

where

$$B_1 \equiv (\beta_1 + \beta_2 + \lambda_1 + \lambda_2) \qquad (11.119)$$

$$B_2 \equiv B_1(\lambda_1\lambda_2 + \mu_1\mu_2 + \lambda_1\mu_2 + \mu_1\lambda_2) + B_3 \qquad (11.120)$$

$$B_3 \equiv \alpha[(\lambda_2 + \beta_1 + \beta_2)(\lambda_1 + \beta_1) + \beta_2^2 + \beta_2\lambda_2] \qquad (11.121)$$

$$P_1 = \frac{[B_1\lambda_1 + \alpha(\lambda_1 + \beta_2)]\beta_2}{B_1(\lambda_1\lambda_2 + \mu_1\mu_2 + \lambda_1\mu_2 + \mu_1\lambda_2) + B_3} \qquad (11.122)$$

$$P_2 = \frac{\beta_1[\lambda_2 B_1 + \alpha(\lambda_2 + \beta_1)]}{B_2} \qquad (11.123)$$

$$P_3 = 1 - \left\{\frac{B_1\beta_1\beta_2}{B_2} + \frac{\alpha[\lambda_2 B_1 + \alpha(\lambda_2 + \beta_1)]}{B_2} + P_2\right\} \qquad (11.124)$$

The steady-state availability, AV_s, is obtained from Eqs. (11.118), (11.122), and (11.123) as follows:

$$\mathrm{AV}_s = P_0 + P_1 + P_2$$

$$= \frac{B_1\beta_1\beta_2 + [B_1\lambda_1 + \alpha(\lambda_1 + \beta_2)]\beta_2 + \alpha[\lambda_2 B_1 + \alpha(\lambda_2 + \beta_1)]}{B_2} \qquad (11.125)$$

Example 11.11 A system consists of two nonidentical, independent, and active units. Whenever a unit fails it is repaired back to its fully operational state. In addition, both units may fail due to common-cause failures. In fact the system follows the same state-space diagram as shown in Fig. 11.9. Unit 1 and unit 2 constant failure rates are $\lambda_1 = 0.1$ and $\lambda_2 = 0.2$ failures/day, respectively. In addition, unit 1 and unit 2 repair rates are $\beta_1 = 0.5$ and $\beta_2 = 1$ repairs/day. The system constant common-cause failure rate is $\alpha = 0.05$ failures/day. Calculate the system steady-state availability.

By utilizing the given data in Eq. (11.118), the steady-state probability of both units operating is

$$P_0 = B_1\beta_1\beta_2/B_2 = (1.85)(0.5)(1)/(1.443)$$
$$= 0.6410$$

Similarly, from Eq. (11.122) for given data, the steady-state probability of unit 2 operating and unit 1 failed is given by

$$P_1 = [B_1\lambda_1 + \alpha(\lambda_1 + \beta_2)]\beta_2/B_2$$
$$= [(1.85)(0.1) + 0.05(0.1 + 1)](1)/(1.443)$$
$$= 0.1663$$

The steady-state probability of unit 1 operating and unit 2 failed from Eq. (11.123) is

$$P_2 = \beta_1[\lambda_2 B_1 + \alpha(\lambda_2 + \beta_1)]/B_2$$
$$= (0.5)[(0.2)(1.85) + (0.05)(0.2 + 0.5)]/1.443$$
$$= 0.1403$$

By summing the above three probabilities, the system steady-state availability, AV_{ss}, is

$$AV_{ss} = P_0 + P_1 + P_2$$
$$= 0.6410 + 0.1663 + 0.1403$$
$$= 0.9476$$

11.5 SUMMARY

This chapter presents the reliability evaluation of commonly known two-state device networks, three-stage device networks and two-state device networks with common-cause failures. A device is said to have two states if it can be either in operating state or in failed state. The networks which consist of such devices are series, parallel, k-out-of-m units, and standby. The reliability and mean time to failure expressions are developed for such configurations.

In addition, this chapter presents derivations of state probability equations for a repairable and a nonrepairable two-failure-mode device with the aid of the Markov technique. Furthermore, reliability expressions are developed for series and parallel networks composed of such devices. In the chapter, the two-failure-mode device is assumed to be an electrical or electronic device. However, the equations developed for series and parallel configurations can also be used (with minor changes) when the system elements are fluid flow valves. Further information on reliability evaluation of networks composed of two-failure-mode devices, the reader should consult Refs. 3 and 4.

One of the areas of current reliability research is the topic of common-cause failures. Therefore, this chapter also covers this area. The reasons for the occurrence of common-cause failures are summarized. The chapter presents mathematical models to evaluate reliability and availability of redundant two-state device systems with common-cause failures. Those readers who wish to delve deeper into the subject of common-cause failures should consult the literature listed in Ref. 8.

EXERCISES

1. A system is composed of seven independent subsystems. All subsystems must function normally for the successful operation of the system. Subsystems' 1–7 constant failure rates are as follows:

   $\lambda_1 = 0.0001$ failures/hr
   $\lambda_2 = 0.0004$ failures/hr
   $\lambda_3 = 0.0002$ failures/hr
   $\lambda_4 = 0.0003$ failures/hr
   $\lambda_5 = 0.0002$ failures/hr
   $\lambda_6 = 0.0007$ failures/hr
   $\lambda_7 = 0.0008$ failures/hr

   Calculate the system reliability for a 20-hr mission.
2. An electronic equipment comprises five active, independent, and identical units. The equipment will only operate successfully if at least three of the units are operating normally. Each unit has a constant failure rate, $\lambda = 0.004$ failures/hr. Calculate the system mean time to failure.
3. Write down time-dependent reliability expressions for the following independent and identical unit networks:
   a. Two (active) units in parallel
   b. Two-unit standby with perfect switching

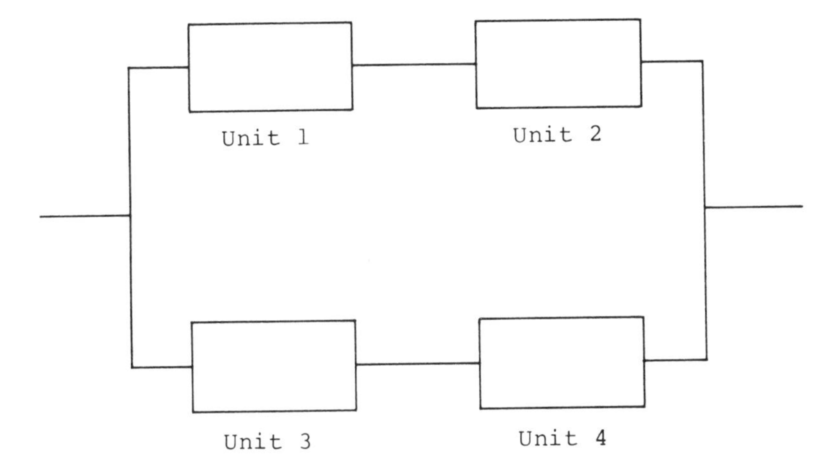

Figure 11.11 A parallel-series network.

Assume each unit failure rate is constant. Compare both network reliability equations. Develop expressions for mean time to failure of these networks and comment on the resulting expressions.

4. Prove that the mean time to failure of the independent and active unit network shown in Fig. 11.11 is given by

$$\text{MTTF} = 0.75\lambda$$

where λ is the unit constant failure rate.

5. List as many as possible of the causes of the occurrence of common-cause failures. Discuss each of these causes in detail by giving real-life situation examples in as many cases as possible.
6. Figure 11.9 shows a state-space diagram of a parallel system. The following additions are proposed to that diagram:
   a. When both units fail normally, the system is repaired back to both unit operating state at a constant repair rate, μ_1.
   b. When both units fail due to a common cause, the system is also repaired back to both unit operating state at a constant repair rate, μ_2.

   Assume both these additions were implemented; develop a steady-state probability expression for the system (if it exists) when both units operate normally by using the Markov technique.
7. Compare the reliability of a two-state device series network with the reliability of a three-state device series network.

REFERENCES

1. W. H. Von Alven, editor, *Reliability Engineering*, Prentice-Hall, Englewood Cliffs, New Jersey, 1964.
2. G. H. Sandler, *System Reliability Engineering*, Prentice-Hall, Englewood Cliffs, New Jersey, 1963.
3. B. S. Dhillon, The Analysis of the Reliability of Multistate Device Networks, Ph.D. dissertation, University of Windsor, available from the National Library of Canada, Ottawa, 1975.
4. B. S. Dhillon, Literature Survey on Three-State Device Reliability Systems, *Microelectron. Reliab.,* Vol. 16 (1977), pp. 601–602.
5. W. C. Ganloff, Common Mode Failure Analysis, *IEEE Trans. Power Appar. Syst.,* Vol. 94 (February 1975), pp. 27–30.
6. J. R. Taylor, A Study of Failure Causes Based on U.S. Power Reactor Abnormal Occurrence Reports, Reliability of Nuclear Power Plants, IAEA-SM-195/16, 1975, pp. 119–130; available from the publishing section, International Atomic Energy Agency, Karntner Ring 11, P.O. Box 590, A-1011 Vienna, Austria.
7. K. N. Fleming, A Redundant Model for Common Mode Failures in Redundant Safety Systems, Proc. Sixth Pittsburgh Annual Modeling and Simulation Conference, 1975, pp. 579–581; available from the Instrument Society of America.
8. B. S. Dhillon, On Common-Cause Failures—Bibliography, *Microelectron. Reliab.,* Vol. 18 (1978), pp. 533–534.
9. B. S. Dhillon, Optimal Maintenance Policy for Systems with Common-Cause Failures, Proc. of the Ninth Pittsburgh Annual Modeling and Simulation Conference, 1978, pp. 1211–1215; available from the Instrument Society of America.
10. B. S. Dhillon, A Common-Cause Failure Availability Model, *Microelectron. Reliab.,* Vol. 17 (1978), pp. 583–584.

12
Reliability Estimation

12.1 INTRODUCTION

Reliability of a piece of equipment is estimated from the failure data. Usually these data are obtained either by testing the equipment during the development phase or from the field use phase. During both these phases special care is given when collecting data. For example, by making sure that the data are obtained under the designed conditions of the equipment. Once the true representative failure data of the equipment are available, the next logical step is to analyze such data. From the failure data one can extract various types of information by analyzing the data from various aspects—for example,

1. The failure probability distribution of the data
2. How well the probability distribution fits the failure data
3. Confidence limits on failure rate, mean time to failure, and reliability
4. Reliability growth of the equipment in question

This chapter describes various aspects associated with reliability estimation in the subsequent sections.

12.2 FAILURE DATA PLOTTING

Once the failure times of items are known, one may utilize such data to develop plots for data hazard rate, data failure density function, data failure distribution function, and data reliability function. According to Ref. 1 the data hazard rate, data failure density function, and data reliability function, respectively, are defined as follows:

$$h(t) = \frac{[\text{number of failures in the time interval}]}{[\text{time interval length}]\left[\begin{array}{l}\text{quantity of survivors at}\\ \text{the starting point of the}\\ \text{time interval}\end{array}\right]}$$

$$= \frac{[k(t_j) - k(t_j + \Delta t_j)]}{[\Delta t_j][k(t_j)]} \quad \text{for} \quad t_j < t \leq t_j + \Delta t_j \qquad (12.1)$$

where

$h(t)$ = the data hazard rate
t = time
$k(t_j)$ = the number of survivors at time t_j
Δt_j = the time interval

$$f(t) = \frac{[\text{number of failures in the time interval}]}{[\text{population size}][\text{time interval length}]}$$

$$= \frac{[k(t_j) - k(t_j + \Delta t_j)]}{[\Delta t_j][K]} \quad \text{for} \quad t_j < t \leq t_j + \Delta_{tj} \qquad (12.2)$$

where $f(t)$ is the data failure density function, K is the size of the original population (i.e., the total number of units put on test at time $t = 0$), and

$$R(t) = \frac{k(t_i)}{K} \qquad (12.3)$$

The failure data plotting is demonstrated with the aid of the above equations in the following example.

Example 12.1 Failure data for 432 identical items is given in Table 12.1. Develop plots for data hazard rate, data failure density function, and data reliability function.

With the aid of Eqs. (12.1)–(12.3) and the data of Table 12.1, the values for data hazard rate, data failure density, and data reliability are tabulated in Table 12.2. The plots of data given in Table 12.2 are shown in Figs. 12.1–12.3.

The plot of data hazard rate in Fig. 12.1 clearly indicates that the hazard rate is increasing with respect to time. On the other hand, the failure density and reliability decrease with respect to time as depicted by the plots of Figs. 12.2 and 12.3, respectively.

Table 12.1 Failure Data for Certain Items

Time interval in 100 hr	Failures in each interval
0–100	121
101–200	80
201–300	70
301–400	63
401–500	30
501–600	25
601–700	21
701–800	10
801–900	7
901–1000	5
	Total 432

Table 12.2 Tabulated Values for Hazard Rate, Failure Density, and Reliability

Time interval in 100 hr	Failures in each interval	Number of survivors at the starting of the time interval	Hazard rate ($\times 10^{-2}$)	Failure density ($\times 10^{-2}$)	Reliability
0–100	121	432	0.281	0.2801	1
101–200	80	311	0.2571	0.1852	0.7199
201–300	70	231	0.303	0.1620	0.5347
301–400	63	161	0.3913	0.1458	0.3727
401–500	30	98	0.3061	0.0694	0.2269
501–600	25	68	0.3676	0.0579	0.1574
601–700	21	43	0.4884	0.0486	0.0995
701–800	10	22	0.4545	0.0231	0.0509
801–900	7	12	0.5833	0.016	0.0278
901–1000	5	5	1	0.011	0.0116

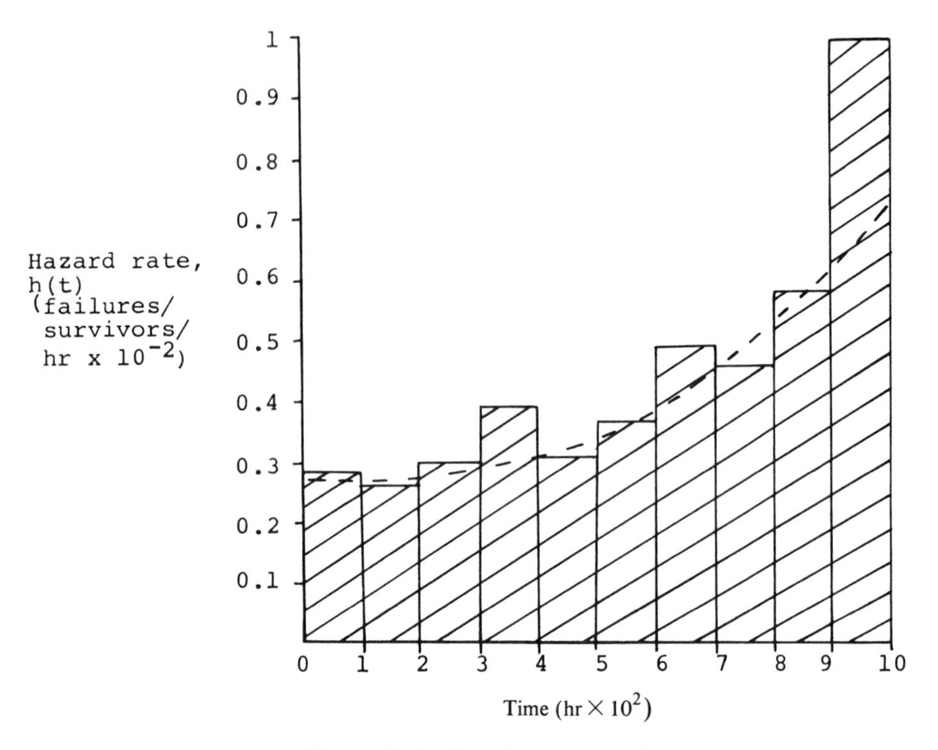

Figure 12.1 Data hazard rate plot.

12.3 RELIABILITY TESTING AND GROWTH MONITORING

This section briefly describes reliability testing and one widely used reliability model to monitor the reliability growth of an item.

12.3.1 Reliability Testing

According to Ref. 2, the reliability tests can be classified into the following three categories:

1. Reliability development tests
2. Reliability demonstration tests
3. Acceptance tests
4. Operational tests
5. Qualification tests

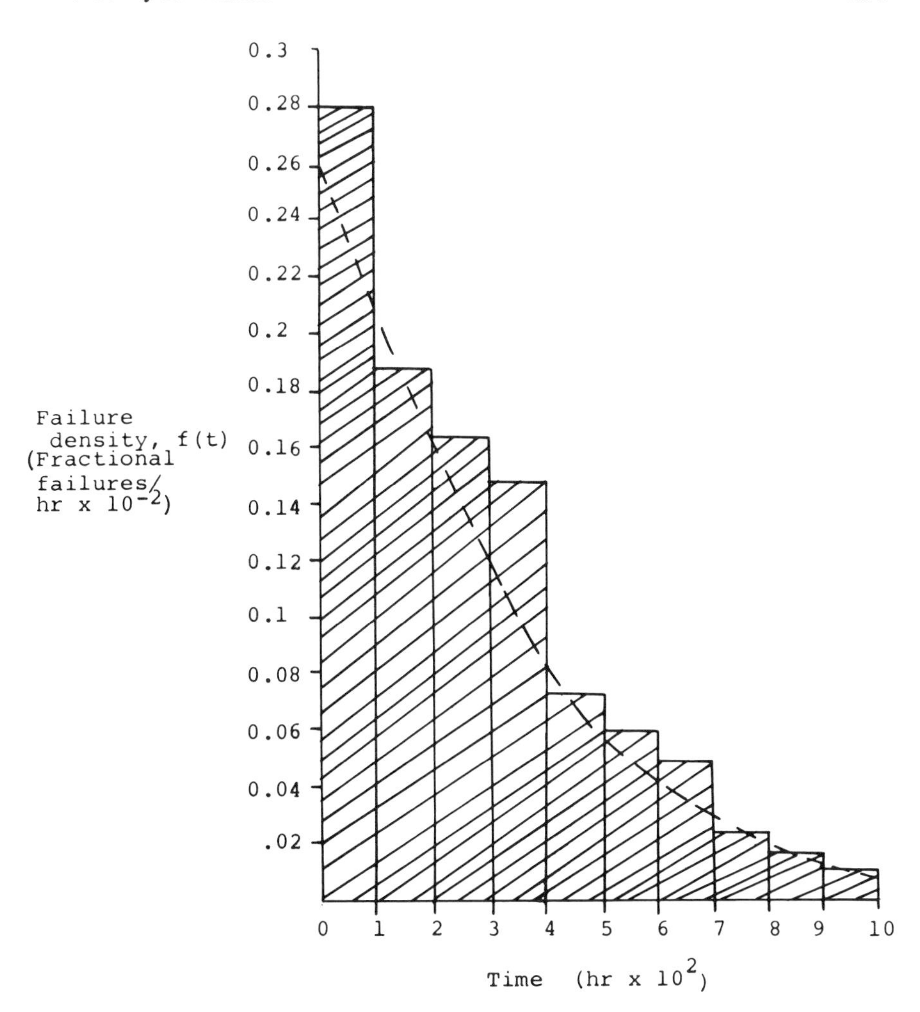

Figure 12.2 Data failure density function plot.

The objectives of tests such as reliability development tests and demonstration tests are to identify any necessary needs to modify product design, to identify the need to improve product design in order to meet the specified product reliability, and to verify product reliability improvements.

Reliability development and demonstration tests are usually directed by the same people who are responsible for the design and development of the product under development. Thus the objectives of these tests are closely

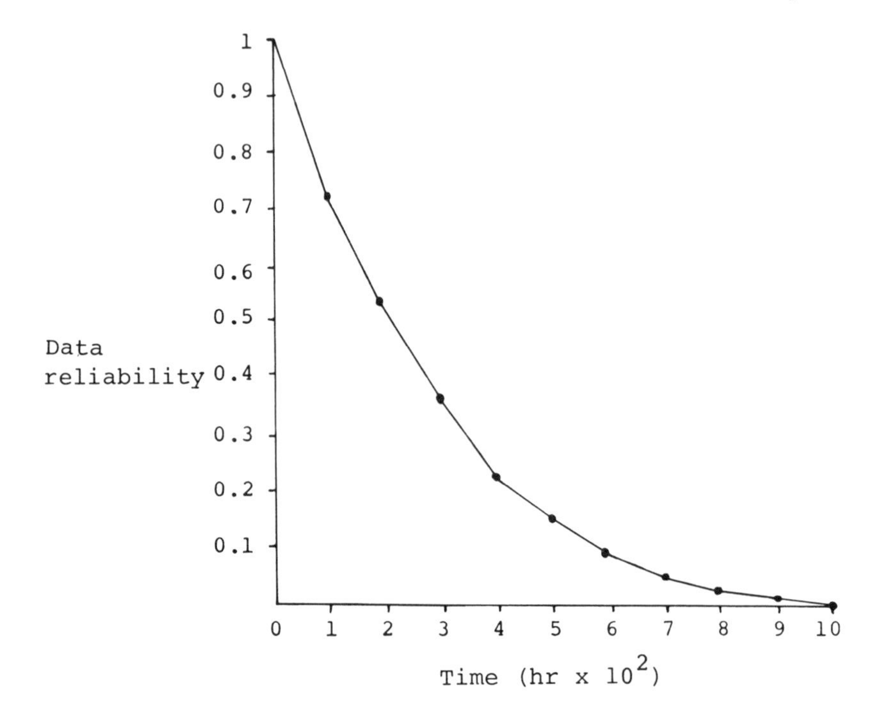

Figure 12.3 Data reliability function plot.

related to the development of the product's prototype. However, these objectives are also applicable to the production phase.

Basically the objectives of qualification tests and acceptance tests are, respectively, to evaluate whether a certain design is suitable (qualified) for its intended use, and to evaluate whether or not an item or a component is acceptable.

There are various objectives of operational testing, for example, verification of predicted reliability and providing data which will be useful in later activities.

12.3.2 Reliability Growth Modeling

During the product development phase, the product reliability grows owing to improvements in design, correcting failures, and so on. To monitor such reliability growth, in 1962, J.T. Duane [3] developed a mathematical model. According to Duane's findings the plot of a product's cumulative mean time between failures against total time yields a straight line on a log–log paper. Thus, according to Ref. 4

$$\log m_c = \log m_s + \beta \log t - \beta \log t_s \tag{12.4}$$

$$m_c = \frac{\text{total time}}{(\text{total number of product failures})} = \frac{t}{k} \tag{12.5}$$

where m_c is the cumulative mean time between failures, m_s is the cumulative mean time between failures at the beginning of the reliability monitoring time period, t_s, and β is the slope parameter (usually it takes values between 0.2 and 0.4 [4]).

From Eq. (12.4), we get

$$m_c = \frac{m_s t^\beta}{t_s^\beta} \tag{12.6}$$

By rewriting Eq. (12.5), we get

$$k = \frac{t}{m_c} \tag{12.7}$$

Substituting Eq. (12.6) into Eq. (12.7) results in

$$k = \frac{tt_s^\beta}{m_s t^\beta} = \frac{t^{1-\beta} t_s^\beta}{m_s} \tag{12.8}$$

Differentiating the above equation with respect to t leads to

$$\frac{dk}{dt} = (1 - \beta)\left(\frac{t_s^\beta}{t^\beta m_s}\right) \tag{12.9}$$

With the aid of Eq. (12.6), the above equation reduces to

$$\frac{dk}{dt} = \frac{1}{m_c} - \frac{\beta}{m_c} \tag{12.10}$$

The left-hand side of Eq. (12.10) is the reciprocal of instantaneous mean time between failure, m_{int}, of the population. Thus we let

$$\frac{dk}{dt} = (m_{\text{int}})^{-1} \tag{12.11}$$

Substituting Eq. (12.11) into Eq. (12.10) leads to

$$\frac{1}{m_{\text{int}}} = \frac{1}{m_c} - \frac{\beta}{m_c} \tag{12.12}$$

Therefore,

$$m_{\text{int}} = \frac{m_c}{1 - \beta} \tag{12.13}$$

Example 12.2 A prototype model of an engineering system was initially tested for a 300-hr period during which 5 failures occurred. The specified mean time between failures of the system is 800 hr. Assume that the value of the Duane model slope parameter β is 0.4. Compute the value of additional system test hours.

Utilizing the specified data in Eq. (12.13) yields value for

$$m_c = m_{\text{int}}(1 - \beta) = 800(1 - 0.4) = 480 \text{ hr}$$

Substituting the given data into Eq. (12.5) yields the estimated value for

$$\hat{m}_s = \frac{300}{5} = 60 \text{ hr}$$

Thus from Eq. (12.6) we get

$$480 = (60)\frac{t^{0.4}}{(300)^{0.4}} = (6.1278)t^{0.4}$$

Therefore,

$$t = \left(\frac{480}{6.1278}\right)^{1/0.4}$$

$$= 54{,}305.8 \text{ hr}$$

$$\text{Additional system test hours} = (54{,}305.8) - (300)$$
$$= 54{,}005.80 \text{ hr}$$

Thus the system has to be tested for another 54,005.80 hr.

12.3.3 Tests for the Validity of Exponentially Distributed Failure Times

In the reliability analysis it is frequently assumed that an item's failure times are exponentially distributed. In order to verify this assumption, various tests are available in the published literature [5–8]. Therefore, this section describes two tests.

Test I

This test is known as the Bartlett test. According to Ref. 5, the Bartlett test statistic is defined as

Table 12.3 Failure Times (in Days)

7	35	85	142
8	46	86	186
20	45	111	185
19	63	112	266
34	64	141	267

$$S_{bk} = 12k^2 \left(\ln X - \frac{Y}{k} \right) / (6k + k + 1) \tag{12.14}$$

$$X \equiv \frac{1}{k} \sum_{i=1}^{k} t_i \tag{12.15}$$

$$Y \equiv \sum_{i=1}^{k} \ln t_i \tag{12.16}$$

where t_i is the ith time to failure and k is the total number of failures in the sample.

A sample of at least 20 failures is necessary for the test to discriminate effectively. If the failure times are exponentially distributed, then s_{bk} is distributed as chi-square with $(k - 1)$ degrees of freedom. Thus, a two-tailed chi-square approach (criterion) is utilized [7].

Example 12.3 A sample of 20 failure times (in days) of an air traffic control system is given in Table 12.3. Determine with the aid of Bartlett's test that the Table 12.3 data are representative of an exponential distribution.

Substituting the specified data into Eq. (12.15) yields

$$X = \frac{1}{20}(7 + 8 + 20 + 19 + 34 + 35 + 46 + 45 + 63 + 64 + 85$$
$$+ 86 + 111 + 112 + 141 + 142 + 186 + 185 + 266 + 267)$$
$$= 96.10$$

Similarly, from Eq. (12.16) we get

$$Y = 82.8311$$

With the aid of the above results from Eq. (12.14) we get

$$s_{b20} = 12(20)^2 \left[\ln(96.10) - \frac{(82.8311)}{(20)} \right] / [6(20) + 20 + 1]$$

$$= 14.43$$

From Table 12.4 for a two-tailed test with 90 percent confidence level, the critical values of

$$\chi^2 \left[\frac{\theta}{2}, (k-1) \right] = \chi^2 \left[\frac{0.1}{2}, (20-1) \right] = 30.14$$

where

$$\theta = 1 - (\text{confidence level}) = 1 - 0.90 = 0.1$$

$$\chi^2 \left[\left(1 - \frac{\theta}{2}\right), (k-1) \right] = \chi^2 \left[\left(1 - \frac{0.1}{2}\right), (20-1) \right] = 10.12$$

Table 12.4 Chi-Square Distribution

Degrees of freedom	Probability			
	0.975	0.95	0.05	0.025
1	0.001	0.004	3.84	5.02
2	0.05	0.1	5.99	7.38
3	0.22	0.35	7.82	9.35
4	0.48	0.71	9.49	11.14
5	0.83	1.15	11.07	12.83
6	1.24	1.64	12.59	14.45
7	1.69	2.17	14.07	16.01
8	2.18	2.73	15.51	17.54
9	2.7	3.33	16.92	19.02
10	3.25	3.94	18.31	20.48
11	3.82	4.58	19.68	21.92
12	4.4	5.23	21.92	23.34
13	5.01	5.89	22.36	24.74
14	5.63	6.57	23.69	26.12
15	6.26	7.26	25.00	27.49
16	6.91	7.96	26.30	28.85
17	7.56	8.67	27.59	30.19
18	8.23	9.39	28.87	31.53
19	8.91	10.12	30.14	32.85
20	9.59	10.85	31.41	34.17
21	12.40	13.85	36.42	39.36

The above results exhibit that there is no contradiction to the assumption of exponential distribution.

Test II

This is another test which is used to determine whether the failure data follow an exponential distribution. This test requires computing the value of c_t, the χ^2 variate with $2k$ degrees of freedom. The value of c_t is given [5,6,9] by

$$c_t = -2\sum_{j=1}^{k} \ln\left[\frac{t(\tau_j)}{t(\tau)}\right] \tag{12.17}$$

where k is the number of failures in the sample, $t(\tau)$ is the total operating time at the termination point of the test, and $t(\tau_j)$ is the total operating time when the jth failure occurs.

If the value of c_t falls within

$$\chi^2\left(\frac{\theta}{2}, 2k\right) < c_t < \chi^2\left(\frac{1-\theta}{2}, 2k\right)$$

then the assumption of an exponential distribution is not contradicted. The θ is the risk of rejecting a true assumption and is given by

$$\theta = 1 - (\text{confidence level})$$

Example 12.4 A sample of 25 identical items were tested for 150 hr, out of which seven items failed. The failed items were not replaced. The times when the items failed are given in Table 12.5. Determine the validity of the exponential distribution.

Table 12.5 Failure Times

Failure number	Failure time (hours)
1	66
2	78
3	26
4	11
5	18
6	40
7	51

$$\text{Total operating time at the termination of test} = (25-7)(150) + (66+78 + 26 + 11 + 18 + 40 + 51) = 2990 \text{ hr}$$

Substituting the specified data into Eq. (12.17) results in

$$c_t = -2\left[\ln\frac{(11)(25)}{(2990)} + \ln\frac{11+(18)(24)}{2990} + \ln\frac{(29)+(26)(23)}{2990} + \ln\frac{(55)+(40)(22)}{2990} + \ln\frac{(95)+(51)(21)}{2990} + \ln\frac{(146)+(66)(20)}{2990} + \ln\frac{(212)+(78)(19)}{2990}\right]$$

$$= -2(-9.2429)$$

$$= 18.49$$

For a 90 percent confidence level we get

$$\theta = 1 - 0.90 = 0.1$$

and

$$\chi^2\left[\frac{0.1}{2}, 2(7)\right], \quad \chi^2\left[\left(1-\frac{0.1}{2}\right), 2(7)\right]$$

Thus from Table 12.4, we have

$$\chi^2(0.05, 14) = 23.69$$

and

$$\chi^2(0.95, 14) = 6.57$$

The value of c_t lies between the above two critical limits, i.e.,

$$6.57 < 18.49 < 23.69$$

In conclusion, the assumption of exponential distribution is not contradicted.

12.4 ESTIMATION OF CONFIDENCE LIMITS

This section is concerned with the estimation of confidence limits on exponential mean life. The chi-square distribution is utilized in establishing the confidence interval limits on mean life.

Usually sampled data are used when estimating the mean life of a product. If one draws two separate samples from a population for the purpose of estimating the mean life, it will be quite unlikely that both samples will yield the same mean life results. Therefore, the confidence limits on mean life are computed to take into consideration the sampling fluctuations. In this section the confidence limit formulations for the following two types of test procedures are presented.

12.4.1 Test Procedure I

In this situation, the items are tested until the preassigned failures occur. Thus from Ref. 2, the formulas for one-sided (lower limit) and two-sided (upper and lower limits) confidence limits, respectively, are as follows:

$$\left[\frac{2t}{\chi^2(\theta,2k)}, \infty\right] \tag{12.18}$$

and

$$\left[\frac{2t}{\chi^2(\theta/2, 2k)}, \frac{2t}{\chi^2(1-\theta/2, 2k)}\right] \tag{12.19}$$

where k is the total number of failures and θ is the probability that the interval will not contain the true value of mean life [thus $\theta = 1 - $ (confidence level)].

The value of t is given by

$$t = xy \quad \text{(for replacement tests, i.e., failed items replaced or repaired)} \tag{12.20}$$

and

$$t = \sum_{j=1}^{k} y_j + (x-k)y \quad \text{(for nonreplacement tests, i.e., failed items are not replaced)} \tag{12.21}$$

where x is the total items, at time zero, placed on test; y is the time at the conclusion of life test; and y_j is the time of failure j.

Example 12.5 A sample of 25 identical electronic components were tested until the occurrence of the twelfth failure. Each failed component was replaced. The last component failure occurred at 150 hr. At 97.5 percent confidence level compute the value of the one-sided (lower) confidence limit (i.e., the minimum value of mean life.)

Substituting the given data into Eq. (12.20) leads to

$$t = (25)(150) = 3{,}750 \text{ hr}$$

The acceptable risk of error is

$$\theta = 1 - (\text{confidence level})$$
$$= 1 - 0.975 = 0.025$$

Hence, with the aid of Eq. (12.18) we get

$$\left\{ \frac{2(3750)}{\chi^2[0.025, (2)(12)]}, \infty \right\} = \left(\frac{7500}{39.36}, \infty \right) = (190.55, \infty)$$

The minimum value of mean life is 190.55 hr for the 97.5 percent confidence level.

12.4.2 Test Procedure II

This is another test procedure in which the testing is terminated at a preassigned number of test hours. From Ref. 2, the formulas for one-sided (lower limit) and two-sided (upper and lower limits) confidence limits, respectively, are as follows:

$$\left[\frac{2t}{\chi^2(\theta, 2k+2)}, \infty \right] \tag{12.22}$$

and

$$\left[\frac{2t}{\chi^2(\theta/2, 2k+2)}, \frac{2t}{\chi^2(1-\theta/2, 2k)} \right] \tag{12.23}$$

The symbols k and θ are defined in the previous section.

Example 12.6 A sample of 25 identical components was drawn from a population and put on test at time $t = 0$. The failed components were not replaced and the test was terminated at 120 hr. Six components failed during the test period at 15, 22, 30, 50, 67, and 85 hr. At 97.5 percent confidence level compute the value of one-sided (lower) confidence limit (i.e., the minimum value of mean life).

By substituting the specified data into Eq. (12.21) we get

$$t = (15 + 22 + 30 + 50 + 67 + 85) + (25 - 6)(120)$$
$$= 2{,}549 \text{ hr}$$

The acceptable risk of error is

$$\theta = 1 - (\text{confidence level})$$
$$= 1 - 0.975 = 0.025$$

With the aid of Eq. (12.22) and the above results, we obtain

$$\left\{ \frac{2(2{,}549)}{\chi^2[0.025,\, 2(6)+2]},\ \infty \right\} = \left[\frac{5{,}098}{\chi^2(0.025,\, 14)},\ \infty \right]$$

$$= \left(\frac{5{,}098}{26.12},\ \infty \right) = (195.18,\ \infty)$$

Thus the minimum value of mean life is 195.18 hr for the 97.5 percent confidence level.

12.5 SUMMARY

This chapter briefly describes the topic of reliability estimation and its closely related aspects. The first topic discussed in the chapter is failure data plotting. The definitions of hazard rate, failure density function, and reliability function are presented. The failure data plotting is demonstrated with the aid of one numerical example. Reliability testing and growth monitoring is the next topic covered in the chapter. Five different types of reliability tests are briefly described. One reliability growth monitoring model known as Duane's model is presented along with one numerical example.

The next topic of the chapter is tests for the validity of the exponentially distributed failure time assumption. Two such tests are described with the aid of two numerical examples.

The last section of the chapter is concerned with estimating the confidence limits on the exponential mean life. Formulas to estimate one-sided and two-sided limits are presented. Application of such formulas is demonstrated with the aid of two numerical examples.

EXERCISES

1. Test failure data of 14 identical mechanical components are given in Table 12.6. Develop plots for data hazard rate, data failure density function, and data reliability function. Comment on the shape of plots.
2. Discuss the following two terms:
   a. Reliability demonstration test
   b. Qualification test
3. A piece of equipment under development was initially tested for 400 hr. Seven failures occurred during this period. The specified value of the equipment mean time between failure is 1000 hr. Calculate the value of additional equipment test hours, if the value of Duane's model slope parameter β is 0.3.

Table 12.6 Failure Data

Failure number	Failure times (hr)
1	10
2	23
3	30
4	40
5	55
6	70
7	85
8	110
9	150
10	180
11	205
12	240
13	280
14	320

Table 12.7 Failure Data

Failure number	Failure times (hr)
1	10
2	15
3	25
4	45
5	60
6	85
7	130
8	160
9	200

4. Twenty identical items were tested for 200 hr. Nine of the total items failed during the test period. Their failure times are specified in Table 12.7. The failed items were never replaced. Determine whether the failure data represent the exponential distribution.
5. In Example 12.6., calculate the values of upper and lower confidence limits on mean life.

REFERENCES

1. M. L. Shooman *Probabilistic Reliability: An Engineering Approach*, McGraw-Hill, New York, 1968.

2. W. H. Von Alven, editor, *Reliability Engineering*, Prentice-Hall, Englewood Cliffs, New Jersey, 1964, p. 508.
3. J. T. Duane, Learning Curve Approach to Reliability Monitoring, *IEEE Trans. Aerospace*, Vol. 2 (1964), pp. 563–566.
4. P. D. T. O'Connor, *Practical Reliability Engineering*, Heyden & Son Ltd., London, 1981, pp. 203–206.
5. B. Epstein, Tests for the Validity of the Assumption that the Underlying Distribution of Life Is Exponential, *Technometrics*, Vol. 2 (1960), pp. 83–101.
6. B. Epstein, Tests for the Validity of the Assumption that the Underlying Distribution of Life Is Exponential, *Technometrics*, Vol. 2 (1960), pp. 167–183.
7. L. R. Lamberson, An Evaluation and Comparison of Some Tests for the Validity of the Assumption that the Underlying Distribution of Life Is Exponential, *AIIE Trans.*, Vol. 12 (1974), pp. 327–335.
8. K. C. Kapur and L. R. Lamberson, *Reliability in Engineering Design*, John Wiley & Sons, New York, 1977, pp. 239.
9. Handbook: Reliability Engineering, NAVWEPS-00-65-502; published by the Direction of the Chief of the Bureau of Naval Weapons, 1964, pp. A3-5 to A3-6; available from the Superintendent of Documents, U.S. Government Printing Office, Washington, D.C. 20402.

13

Introduction to Engineering Design

13.1 INTRODUCTION

All the engineering products found in the market today were designed by engineering experts in the past. When these products are used in the field, subject to the designed conditions, their performance reflects the quality of their design. Obviously, the poorly designed products will have poor performance in the field. There are various factors which play an important role in achieving the best design at the minimum cost. In Ref. 1, engineering design is defined as the activity where various methods and scientific principles are used to decide the selection of materials and the placement of these materials to develop an item which fulfills specified requirements.

There are various reasons for designing and purposes of design. Some of them are as follows [2, 3]:

1. Reducing the cost
2. Developing a new way
3. Producing a useful item
4. Producing a physically realizable product
5. Lowering hazard
6. Reducing inconvenience
7. Producing an item with economic worth
8. Meeting competition
9. Developing the market
10. Meeting social changes

This chapter briefly describes the various important aspects of engineering design.

13.2 TYPES OF DESIGN

There are various specialized classifications of design [4]. Figure 13.1 shows five different types of design. These are as follows:

1. Engineering design
2. Industrial design
3. Process design
4. Visual design
5. Product design

Engineering design is concerned with applying various techniques and scientific principles to the development and analysis of basic functional features for systems, devices, etc. Another type of design is known as industrial design. This is that type of design which usually designates an independent design effort. In other words, it is not a part of a specific product manufacturing organization. Usually an individual with combined abilities in areas such as product design, styling, and engineering performs this type of design activity as a consultant to product-producing organizations.

Process design is another type of design. That is usually concerned with that type of design which is restricted to the design of components, tools, equipment, etc., in other words, those items which are for mass production systems.

Visual design is also known as the styling design. This type of design is usually concerned with the appearance features of an item. Lastly product design is normally associated with specifically those items which are ultimately to be sold to consumers.

13.3 FUNCTIONS ASSOCIATED WITH ENGINEERING DESIGN

There are various functions involved in engineering design. In Ref. 5, they are classified into five broad categories. These are as follows:

1. Manufacturing functions
2. Commercial functions
3. Engineering functions
4. Quality assurance functions
5. Research functions

All these classifications are described below:

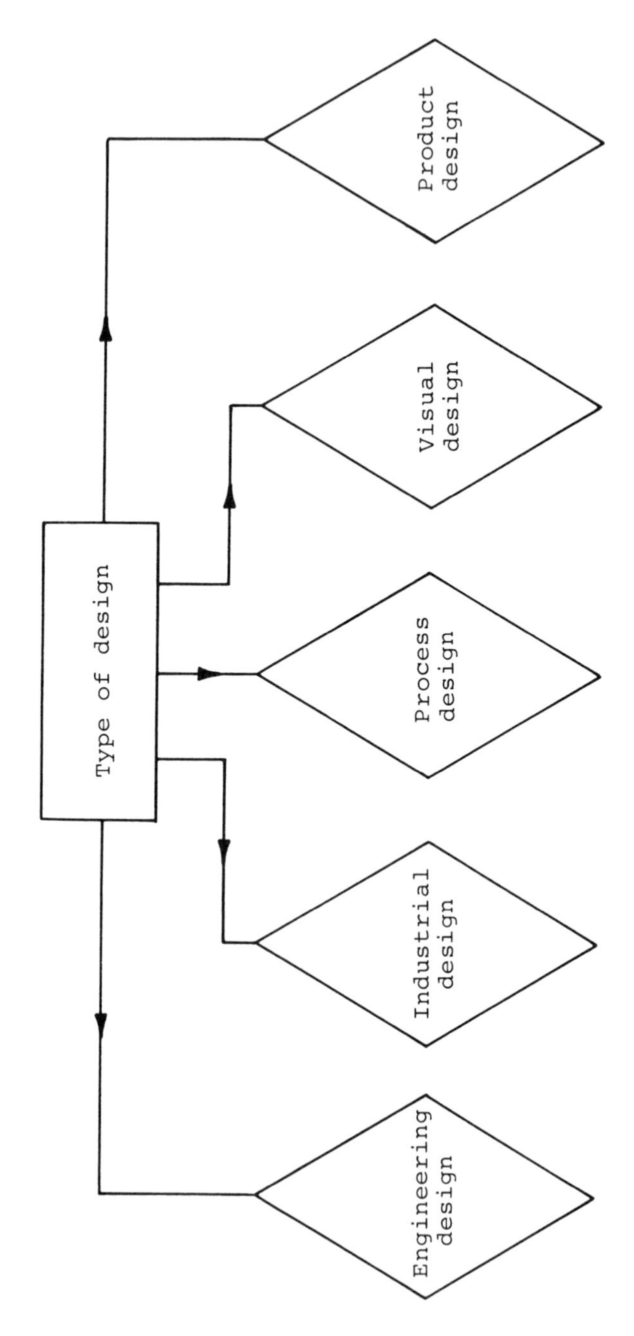

Figure 13.1 Types of design.

Manufacturing Functions

Manufacturing functions include all those functions which are concerned with manufacturing. Examples of such functions are as follows:

1. Assembly
2. Finding out the tooling requirement
3. Manufacturing planning
4. The design of tools
5. Detail manufacture
6. Keeping pace with the latest manufacturing methods
7. Purchasing materials
8. Cost control

Commercial Functions

Commercial functions involve relationships with various clients. Some of these functions are as follows:

1. Conducting market surveys and tendering
2. Managing contracts effectively
3. Arranging delivery
4. Advertising the company and its products
5. Arranging payment

Engineering Functions

Engineering functions are subcomponents of the design activity. Examples of such functions are as follows:

1. Developing new design concepts
2. Designing for production
3. Supporting functions: estimating cost, analyzing field problems, the provision of maintenance instructions, etc.

Quality-Related Functions

Quality-related functions are concerned with the quality of the end product. These functions are relevant to areas such as the following:

1. Design methods and procedures
2. Design auditing setup
3. Quality and design data

Research Functions

These functions are associated with research. Examples of these function are are as follows:

1. Conducting basic and applied research
2. Preparing specifications for quality testing procedures
3. Preparing process specifications for welding
4. Preparing process specifications for the testing of highly stressed parts

13.4 CHARACTERISTICS OF GOOD AND BAD DESIGNS

This section lists, separately, the characteristics of good and bad designs [3].

Good Design

Some of its characteristics are as follows:

1. Reliability
2. Long useful life
3. Low maintenance and noise level (if any)
4. High accuracy
5. Low cost
6. Attractive appearance
7. Trouble-free
8. Simplicity

Cheap or Bad Product Design

Following are some of the characteristics of cheap or bad product design:

1. Poor accuracy
2. High noise level and nonadjustable
3. Poor reliability
4. Nonrepairable and flimsy
5. Wears out
6. Rattles and rusts
7. Cracks and corrodes
8. High maintenance cost
9. Short useful life

13.5 DESIGN PROCEDURE

This section briefly describes a seven-step procedure used for designing [6]. These steps are listed below:

Step 1: Perform analysis of requirements.
Step 2: Define the scope, objectives, and pertinent restraints with respect to the requirements; identify any significant problem which has to be solved.
Step 3: Develop alternative solutions.
Step 4: Perform feasibility analysis of alternative solutions.
Step 5: Optimize the promising solutions.
Step 6: Select the solution for use.
Step 7: Implement the solution.

The design process is described in detail in the next chapter.

13.6 ARGUMENTS FOR AND AGAINST FINITE-LIFE DESIGN

This section presents the arguments for and against finite-life design taken from Ref. 7. This reference presents the result of a survey conducted among 1000 design engineers and engineering supervisors. Some of their responses for and against finite-life design were as follows:

For

1. It is useful because high product turnover creates more jobs, lower cost, and so on.
2. It is helpful for innovation and design development.
3. It is useful to design finite-life product because, due to technical progress, the very-long-life product may have to be scrapped.
4. It is not wise to build very-long-life products, if their life is not needed.
5. A very-long-life product cannot be afforded by any one.

Against

1. Good design work is discouraged.
2. It degrades the image of the product.
3. Research and development for innovative products is discouraged.
4. Customers expect high quality.
5. It wastes resources and is immoral.
6. It burdens consumers with cost of maintenance.

13.7 CHARACTERISTICS OF ENGINEERS AND RAW MATERIAL AVAILABLE TO THEM

This section briefly lists separately the characteristics of an engineer and raw material available to him or her to accomplish design.

Characteristics of a Typical Engineer

1. Works hard
2. Conservative
3. Primarily interested in mechanical-oriented hobbies
4. Family oriented
5. Takes interest in gardening
6. Reserved and independent
7. Honest and sincere
8. Basically inclined toward casual relationships with colleagues
9. Well motivated and organized
10. Cautious and tends to be conventional
11. Does not appreciate chit-chat
12. Possesses a limited number of close friends
13. Poor communicator
14. Possesses high visual ability
15. Does little reading outside his or her field and has limited cultural interests
16. Basically a shy person
17. Possesses little interest in people
18. Belongs to the top ten percent of the population in regard to intelligence
19. Emotionless
20. Friendly

Raw Materials Available to an Engineer

Raw materials available to an engineer with respect to design can be classified into five categories [10]. These are as follows:

1. Engineering technology
2. Mathematics
3. Natural sciences
4. Engineering sciences
5. Miscellaneous

The engineering technology and sciences include, respectively, areas such as manufacturing methods, experimentation methods, experience, manipulations; and electrical theory, fluid and solid mechanics, material sciences, and thermodynamics. The natural sciences include life and space sciences, earth sciences, physics, chemistry, etc.

The miscellaneous group includes areas such as economics, information theory, psychology, literature, and communications.

13.8 CHARACTERISTICS OF A TYPICAL DESIGN ENGINEER

These characteristics include the following [3]:

1. Very creative
2. Possesses a strong background in basics
3. Basically an observant and curious person
4. Possesses self-confidence
5. Tolerates criticism
6. Possesses high initiative
7. Flexible and persistent
8. Takes certain calculated risks
9. Possesses significant interest in areas such as science and mechanics
10. Possesses a broad intellectual scope
11. Follows creative nonconventional procedure
12. Open minded with respect to experience
13. Does not possess faith in routine
14. Possesses a strong memory
15. Has speaking and writing skills
16. Has visual capacity
17. Possesses capacity to integrate
18. Has ability to think logically
19. Possesses ability to concentrate

13.9 CREATIVITY IN ENGINEERING DESIGN

Creativity plays an important role in engineering design. Thus, this section briefly examines the characteristics of a creative and uncreative person, the sources of design project ideas, and reasons for prevention of innovations. All these areas are described below [3]:

Characteristics of a Creative Person

The characteristics of a creative person include the following [11, 12]:

1. Places no value on job security
2. Possesses good sense of humor

3. Is open to experience
4. Accepts failure easily
5. Is independent and is insensitive to the feelings of other people
6. Is observant and takes interest in exploring ideas
7. Works relentlessly
8. Does not give any importance to status symbols
9. Likes to seek privacy and autonomy
10. Is a nonconformist and accepts chaos
11. Possesses an IQ between 100 and 140
12. Possesses listening ability

Characteristics of a Noncreative Person

Some of the characteristics of a noncreative person are as follows [3]:

1. Fears ridicule and failure
2. Does not believe in nonconventional ideas
3. Is jealous of competitiveness
4. Resists change
5. Is a conformist
6. Looks for security
7. Is cynical
8. Has no interest in experimenting
9. Is inclined towards systematic routine

Design Project Idea Sources

This section presents sources of design project ideas. Some of them are as follows:

1. Consulting companies, journals, reviews, and newspapers
2. Sales and research groups
3. Advertising agencies and procurement agents
4. Trade associations and patents
5. Industrial development agencies and backyard inventors

Reasons for Prevention of Innovations

For mass-produced products some of the reasons for delay or prevention of innovations are as follows:

1. Poor competition
2. Resistance from consumers
3. Unavailability of capital
4. Present commitments of the organization

5. Sociological implications such as overhauling regulatory laws, codes for safety, etc.

All of these reasons are self-explanatory; therefore, they are not described in detail. However, if necessary, their description may be found in Ref. 3.

13.10 SUMMARY

This chapter briefly presents the various aspects of engineering design. Various purposes of design are listed. Five types of design are described. The functions involved in engineering design are explained in detail. These are manufacturing, engineering, commercial quality assurance, and research functions.

Characteristics of good and bad design are presented along with a procedure used for designing. Reasons for and against finite-life design are outlined. This material is the result of a survey conducted among design engineers and engineering supervisors.

Twenty characteristics of a typical engineer are presented along with raw material to accomplish design. In addition, most of the characteristics of a typical design engineer are listed.

The last item covered in the chapter is concerned with creativity in engineering design. The characteristics of a creative person are presented. Also, the characteristics of an uncreative person are listed. Sources of design project ideas and reasons for prevention of innovations are given:

EXERCISES

1. Discuss in detail at least five purposes of design.
2. Describe a systematic procedure used for designing a product.
3. What are the characteristics of a bad design?
4. What are the functions of a design engineer?
5. Discuss the attributes of an engineering designer.
6. What are the factors which prevent innovation?
7. What are the sources which will be helpful to get a new idea for design?

REFERENCES

1. W. H. Middendorf, *Engineering Design*, Allyn and Bacon, Inc., Boston, 1969, p. 2.
2. M. Farr, *Design Management*, Cambridge University Press, London, 1955, pp. 42–59.

3. L. Harrisberger, *Engineersmanship: A Philosophy of Design*, Wadsworth Publishing Company, Belmont, California, 1966.
4. D. H. Edel, *Introduction to Creative Design*, Prentice-Hall, Englewood Cliffs, New Jersey, 1967, pp. 7–8.
5. C. H. Flurscheim, *Engineering Design Interfaces*, Design Council Publication, London, 1977, pp. 12–14.
6. S. F. Love, Design Methodology, *Des. Eng.* (*Toronto*), April (1969), pp. 30–32.
7. E. Raudsepp, The Engineer—Paragon or Paradox? His Personality, *Mach. Des.* December (1959), pp. 24–28.
8. E. Raudsepp, The Engineer—His Intelligence and Abilities, *Mach. Des.*, December (1959), pp. 29–31.
9. E. Raudsepp, The Engineer—His Interests, *Mach. Des.*, January (1960), pp. 25–28.
10. J. P. Vidosinc, *Elements of Design Engineering*, The Ronald Press Company, New York, 1969, pp. 7–8.
11. J. H. McPherson, Are You Creative? *Prod. Eng.* (*N.Y.*), November (1958), p. 28.
12. J. H. McPherson, The Relationship of the Individual to the Creative Process in the Management Environment, American Society for Mechanical Engineers, 1964, Paper 64 MD 12.

14

The Design Process and Associated Areas

14.1 INTRODUCTION

This chapter describes the design process and the areas directly or indirectly associated with the process. The design process may be described as an integration of technical know-how obtained from the laboratory and put to use for both client or customer and marketplace [1]. Furthermore, it is added that the design process transforms ideas into profitable products and information into decisions. The areas of knowledge on which the engineering design process is dependent are "marketology," engineering technology, and scientific related information. The term "marketology" is defined as the business of making design-influencing decisions associated with distribution, consumer reaction, material availability, and so on. Thus, the design process may also be described as an imaginative integration of marketology, engineering technology, and scientific related information toward the development of a profitable product.

Various steps are associated with the design process. According to Ref. 2, some people have outlined the process in as many as 25 steps but others in as few as five steps. For example in Ref. 3 it is outlined in eight steps (namely, need recognition, problem definition, preparation, conceptualization, the design synthesis, evaluation, optimization, and presentation), whereas in Ref. 2 it is composed of six steps (namely, requirement recognition, problem definition, information collection, conceptualization, evaluation, and communication of final design to concerned bodies).

This chapter presents the design process and various related areas in the subsequent sections.

14.2 THE DESIGN PROCESS

This section describes the stages associated with the design process. These are as follows [4]:

1. Problem identification
2. Problem definition
3. Information collection
4. Task specifications
5. Idea generation
6. Conceptualization
7. Analytical stage
8. Experimental stage
9. Solution presentation
10. Production
11. Product distribution
12. Consumption stage

Each of the above stages is described below.

Problem Identification

This is the first stage of the design process and is concerned with a thorough investigation of what the problem or requirement really is. Requirement or need may be generated, for example, by

1. Purchasing representatives
2. Marketing agents
3. Customers of the product
4. Servicing people
5. Operators
6. Trade associations
7. Government bodies

Usually the cause of the need is dissatisfaction with the present condition. Examples of the need are as follows:

1. Improve reliability of the product
2. Reduce cost of the product
3. Improve performance and efficiency of the product

Problem Definition

Establishing the definition of the problem is probably the most critical stage of the design process. Thus special care must be given when defining the problem; otherwise poor definition of the problem may lead to an undesirable end product. According to Ref. [2], it is usually beneficial to establish a broad definition of the problem, if possible. This way unconventional or other solutions to the problem will not be overlooked or at least the chances will be reduced significantly. Questions such as the following [5] are to be asked during this stage:

1. Is the problem under consideration too big?
2. What are the design constraints?
3. What are the objectives which are to be fulfilled by the design?
4. What are the social consequences of the design in question?
5. What should be contained in the definition of the problem?
6. What are the special difficulties associated with the design under consideration?

Information Collection

This stage of the design process is concerned with collecting the necessary design information. The following are the important sources of information:

1. Research documents or reports
2. Patents and journals
3. Component suppliers catalogs
4. Handbooks
5. Abstracts
6. Suppliers
7. Technical libraries
8. Professional societies and trade associations

Task Specifications

Task specification is concerned with writing down all essential parameters and data tending to control the design under consideration and direct it toward the set objective. The objective of task specification is to help the designer toward the set goal. The written material is concerned with maintenance and cost and takes into consideration the design development environment, that environment under which the design is to be developed.

Idea Generation

This is concerned with the generation of new ideas. There are various techniques used to stimulate ideas. Examples of such techniques are brainstorming [6], attribute listing, synectics, and checklists. The checklists and brainstorming techniques are described below.

Checklists. This is one of the simplest methods for an individual person to generate new ideas [7]. This technique applies a list of general questions to the problem. The technique assumes the existence of an idea for the solution or the solution itself. A list of typical questions is as follows:

1. Can the existing solution be modified?
2. What are the ways in which the appearance, quality, and performance of the existing solution or idea can be improved?
3. What is the scientific basis for the idea under consideration?
4. What are other scientific bases which will be equally good for the idea in question?
5. Is it possible to combine the solution with another idea?
6. What are other uses of the solution?
7. What are the most and least useful features of the solution under consideration?
8. What are the benefits and drawbacks of the solution?
9. Is it possible to improve benefits of the solution further?
10. Is it possible to overcome drawbacks of the solution?

Brainstorming method. In the modern context, this method was first applied by Alex Osborn [6] in 1938. The term *brainstorming* may simply be described as using the brain to storm a problem. A number of persons with diverse backgrounds but similar interests participate in the brainstorming sessions. Usually the best result is achieved with 8 to 12 persons in each brainstorming session. During the session one solution idea triggers another one and the process continues. The brainstorming session is always less than an hour but sometimes as short as 15 minutes. When conducting a brainstorming session one must follow the guidelines given below:

1. Aim to get at least 50 ideas during each brainstorming session.
2. Do not allow criticism of anybody's idea during the session, no matter how wild it is.
3. Welcome free-wheeling participation during the session.
4. At the end of the session combine and improve ideas.
5. Make sure that the rank of participants is fairly equal. Otherwise, it will take lot of warming up time for lower-rank people to mix freely with higher-rank ones.
6. Make sure the ideas are recorded during the session.

Conceptualization

Conceptualization is a creative and innovative activity in the form of the production of a series of alternative answers to the stated objective. Usually, this activity takes the form of free-hand drawings or sketches.

Analytical Stage

The analytical stage is concerned with testing those concepts against physical laws which have been chosen because they define possible solutions to the specified objective.

Experimental Stage

During this stage of the design process a unit of hardware is built and tested to determine reliability, performance characteristics, workability, and so on. At this stage, the following hardware construction techniques are available to the designer:

1. Prototype
2. Mock-up approach
3. Models

The prototype is the actual physical system which is constructed according to the design outlined on paper. It generates the greatest quantity of useful data but is the most expensive and time-consuming approach. The prototype generates useful information on areas such as the following:

1. Durability
2. Workability
3. Assembly methods
4. Performance subject to real environments

The second type of construction technique available to a designer is known as the mock-up. The mock-up is built to give the designer a feel for his or her design. In addition, it is usually built to scale and the materials used for its construction are wood, cardboard, plastic, and so on. The mock-up technique is the cheapest and simplest to produce, but it provides the least information. The following are some of the main uses of the mock-up approach:

1. To check techniques to be used for assembling
2. To sell the design idea to customers and management
3. To check clearances
4. To check appearance

The third type of construction technique is the model. In this case, four different kind of models are used to represent the physical system. Thus the desired information is collected through these models.

These models are as follows:

1. *Distorted model*: As its name implies, this type of model purposely violates a certain number of design conditions.

2. *True model*: The true model is constructed to scale and is an exact geometric reproduction of the actual system. In addition, it fulfills, all the design parameter requirements.
3. *Adequate model*: This kind of model does not yield any information regarding the entire design. However, it is used to test certain design characteristics.
4. *Dissimilar model*: This model does not resemble the actual system under consideration, but through appropriate analogies it provides useful and accurate behavioral characteristics data.

Solution Presentation

This phase of the design process is concerned with writing the report on design, so the design can easily be communicated to management, client, shop floor, etc. The report should contain items such as the following:

1. Description of the product or equipment in question
2. Description of requirements fulfilled by the proposed product
3. Detailed description of how the proposed product works
4. Construction specifications and assembly drawings
5. Description of cost
6. Description and listing of standard parts

Production, Product Distribution, and Consumption

During the production stage of the design process special care is given to areas such as quality assurance, work scheduling, personnel training, production volume, and availability of facilities. However, during the distribution stage of the design process attention is given to things such as advertising, pricing, best time to release product, and conducting market test. Usually during the production phase, the designer is assisted by specialists in the area of concern. Similarly, usually the product distribution stage is handled by the experts in the area.

Consumption is the final stage of the design process. Basically, this phase is concerned with collecting and analyzing data associated with consumption—for example, performance and reliability of the product in the field, users of the product, reaction of competitors, and other feedback (data) from the consumers. These kinds of data will be very useful to design new products or suitably modify the existing one.

14.3 FACTORS FOR NEW PRODUCT FAILURE

This section lists the factors detrimental to the existence of new products [1]. Thus, the decision taken in the design process must take these factors into consideration. Some of these factors are as follows:

1. Change in preference of consumers
2. Cheaper product of foreign competitors
3. Inaccuracy in the market survey
4. Quick improvement in the product of the competitor
5. Entire market captured because of research "breakthrough"
6. Increase in the cost of production
7. Product market eliminated because of new technological developments
8. A design shortcoming not apparent until the product is in the market
9. Obsolescence of the product because of the pace of technological developments

14.4 DESIGN MORPHOLOGY

The typical design project must be carried through a series of phases in order to obtain the end product. Thus, according to Dieter in Ref. 2 the project will break into various time phases. From Ref. 8, the time phases of the design project are shown in Fig. 14.1.

All the phases outlined in Fig. 14.1 are considered to be self-explanatory; therefore, no attempt is made to describe them here. However, their description may be found in Ref. 2.

14.5 SPECIFICATIONS

Specifications are used when designing a new product. Moreover, the necessary information on the product to be designed is transmitted to the design engineer and others through the specifications. Thus careful consideration must be given when writing such specifications. Some useful hints for developing good specifications for a product are as follows:

1. Make certain that the specification is reasonable for stated tolerances.
2. Minimize reference to standard documents.
3. Ascertain that the specification is concise and complete.
4. Omit ambiguous phrases in the specification.
5. Make use of plain English.
6. Ascertain that the specification is accurate.
7. Make sure that the specification is flexible enough so that it can incorporate further improvements without any difficulty.

Information on items such as follows are to be included in the engineering design specification [9]:

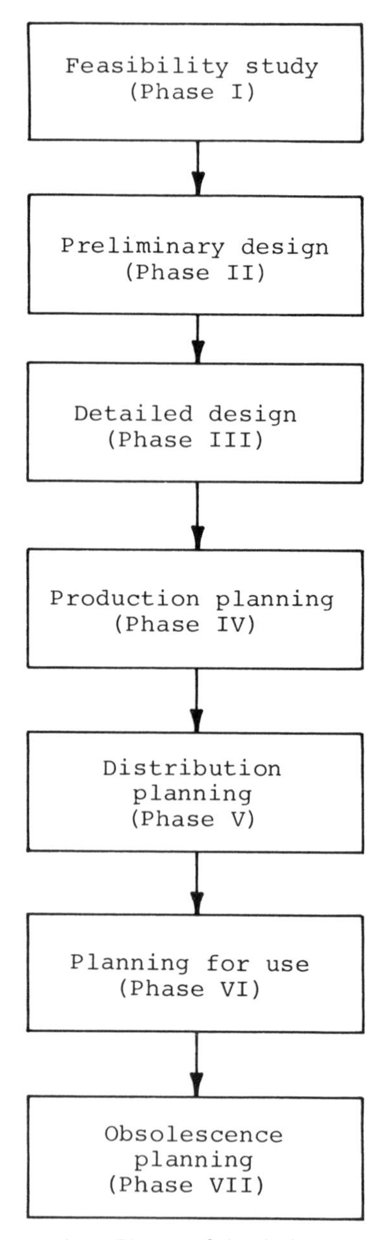

Figure 14.1 Phases of the design product.

1. Limitations due to installation
2. Manufacturing constraints
3. Objective
4. Operation constraints
5. Effect on other systems, subsystems, etc.
6. Functional requirements
7. Environments

All of the above items are described in Ref. [10].

14.6 SUMMARY

This chapter describes the design process and associated areas. Thus the topics covered in the chapter are the design process, factors for new product failure, design morphology, and specifications.

The first topic described in the chapter is the design process. Twelve stages of the process are presented. These stages are problem identification, problem definition, information collection, task specifications, ideas generation, conceptualization, analysis, experimental stage, solution presentation, production, product distribution, and consumption. All of these stages are described in detail.

The next topic covered in the chapter is factors for new product failure. Nine of these factors are presented. Morphology of design phases are shown in Fig. 14.1. These phases are feasibility study, preliminary design, detail design, production planning, distribution planning, planning for use, and obsolescence planning. The last topic of the chapter is specifications. Useful hints to develop good specifications are outlined and items are listed on which information should be included in the engineering design specification.

EXERCISES

1. Describe the meaning of the term "design process."
2. Describe the group brainstorming method and its nonmodern origin.
3. List a number of sources to collect design-related information.
4. Discuss the following two techniques of idea generation:
   a. Synectics
   b. Attribute listing
5. Describe in detail the following terms associated with the design process:
   a. Dissimilar model
   b. True model
   c. Distorted model

    d. Adequate model
6. Explain the meaning of the term "morphology of design."
7. List and discuss at least ten useful guidelines to be considered when writing a specification.
8. What are the benefits and drawbacks of engineering specifications?

REFERENCES

1. L. Harrisberger, *Engineersmanship: A Philosophy of Design*, Brooks/Cole Publishing Company, Belmont, California, 1966, pp. 69–88.
2. G. E. Dieter, *Engineering Design: A Materials and Processing Approach*, McGraw-Hill, New York, 1983, pp. 29–39.
3. J. P. Vidosic, *Elements of Design Engineering*, The Ronald Press Company, New York, 1969, pp. 16–25.
4. P. H. Hill, *The Science of Engineering Design*, Holt, Rinehart and Winston, New York, 1970, pp. 33–60.
5. S. F. Love, Design Methodology: Defining the Problem, *Des. Eng.* (*Toronto*), July (1969), pp. 69–72.
6. A. F. Osborn, *Applied Imagination*, Scribner's, New York, 1963, pp. 151.
7. G. C. Beakley and E. G. Chilton, *Introduction to Engineering Design and Graphics*, Macmillan, New York, 1973, p. 207.
8. M. Asimow, *Introduction to Design*, Prentice-Hall, Englewood Cliffs, New Jersey, 1962, pp. 14.
9. D. J. Leech, *Management of Engineering Design*, John Wiley & Sons, New York, 1972.

15

Engineering Design Reviews

15.1 INTRODUCTION

During the design phase of a product various design review are conducted. The basic objective of design reviews is to assure the application of proper design principles. In addition, the design reviews are conducted to determine whether the design work is progressing according to plans, specifications, and so on. Design reviews are useful to assure reliable, maintainable, and reproducible products as well as helping to reduce design shortcomings. For accelerating design maturity of a product, a properly established engineering design review program serves as a useful tool of the product manufacturer. Furthermore, such a program creates an opportunity to apply a level of engineering talent to the product design; otherwise, it would not be feasible. The design review approach assembles together a team of experienced persons to examine the design in question from all sides, and utilizes the expertise of the team members for the design under study. According to Ref. 1, the budgeted cost for design reviews usually varies between 1 and 2 percent of the total engineering cost on a project.

This chapter dwells on the various aspects of design reviews.

15.2 TYPES OF DESIGN REVIEWS

Different authors and practitioners have classified design reviews differently. However, for our purpose we have classified the design reviews into the following three categories:

a. *Preliminary design review*: This review should be conducted before the formulation of the initial design. The purpose of this type of design review is to examine each requirement outlined in the specification for such things as completeness, validity, and accuracy. According to Ref. 2, items such as the following are to be reviewed as necessary during the preliminary design review:

1. Availability of materials
2. Critical parts
3. Pertinent legislation concerning product design
4. Product liability
5. User or customer
6. Cost objective
7. Requirements of the customer from the product
8. Government and other standards concerning quality, reliability, safety, etc.
9. Constraints associated with the design under study
10. Important data associated with the life cycle of earlier similar products
11. Functions required from the product under review
12. Requirements imposed by schedule
13. Documentation requirement, value engineering proposals, test considerations, and risk
14. Considerations for "make or buy" and design alternatives

b. *Intermediate design review*: This type of design review is conducted before starting the detailed production drawings. However, prior to this review the design selection process is over and preliminary layouts are finished. During the intermediate design review each requirement of the specification is compared with the design. At this stage any design changes can be carried through effectively.

c. *Critical design review*: This is also known as the final design review. It is conducted after the completion of production drawings. During this review a considerable amount of information is available to the design review team, for example test data, reports of earlier design reviews, and cost. Furthermore, during the review emphasis is given to design producibility, value engineering, review of analysis results, and so on.

15.3 SUBJECTS DISCUSSED DURING DESIGN REVIEWS AND ITEMS REQUIRED FOR REVIEWS

This section lists separately the major subjects discussed during design reviews as well as items required for reviews. Both these items are discussed below.

15.3.1 Subjects Discussed During Design Reviews

Usually the major subjects discussed during design reviews are as follows [3]:

1. *Electrical*: Simplification of design, performance, electrical interference, analysis of circuitry, testing, results of testing, and so on.
2. *Mechanical*: Results of tests, balance, connectors, thermal analysis, and so on.
3. *Specifications*: Adherence to specifications, correctness of specifications, etc.
4. *Standardization*: This is concerned with various aspects of standardization.
5. *Reliability*: Reliability allocation, reliability evaluation, redundancy, failure modes and effect analysis, reliability testing, results of reliability studies, parts selection, etc.
6. *Drafting*: Dimensions, tolerances, completeness of notes, partmarking, etc.
7. *Human engineering*: Controls and displays, labeling and marking, glare, seating arrangement (if applicable), etc.
8. *Reproducibility*: Economical assembly of product in the production shop, reliance of products on a single part supplier, etc.
9. *Maintainability*: Interchangeability, meeting minimum downtime, maintenance philosophy, etc.
10. *Value engineering*.
11. *Safety*.
12. *Finishing*.

15.3.2 Items Required for Design Reviews

The design review committee members should have access to items such as the following:

1. Specifications and schematic diagrams
2. Description of circuits
3. List of parts
4. Procedures for testing
5. Circuit performance
6. Inputs and outputs lists
7. Vibration and shock tests data
8. Acceleration and thermal tests data
9. Acoustic noise tests data
10. Reliability predictions
11. Result of failure modes and effects analysis

12. Reliability trade-off studies
13. Reliability evaluation methods
14. Drawings

15.4 DESIGN REVIEW BOARD CHAIRMAN

This section covers the qualities and duties of design review board chairman and useful lead-in items to the design review. The design review chairman should not be in direct line of authority to the designer whose design is to be reviewed. However, the chairman should belong to the engineering department.

15.4.1 Qualities of the Design Review Board Chairman

Some of the qualities of the board chairman are as follows:

1. Has a broad understanding of the technical problem
2. Possesses skill to lead a technical meeting
3. Possesses a high degree of tact
4. Possesses a high degree of discretion
5. Has no inclination for or against the proposed design
6. Possesses a pleasant personality

Position and the general technical competence of the potential chairman also play an important role in the selection process. According to Ref. 4, the configuration manager frequently heads the design review board.

15.4.2 Functions of the Design Review Board Chairman

Some of the functions of the board chairman are as follows:

1. Schedules the reviews as soon as possible
2. Establishes the procedure for choosing specific items for review
3. Chairs the meeting of the design review board
4. Supervises the publication of minutes and looks after their circulation to appropriate bodies
5. Evaluates comments from the design review meetings and directs necessary follow-up action or actions
6. Determines and establishes the type of design review to be conducted
7. Looks after the circulation of copies of the agenda and other materials to concerned bodies well in advance of each review
8. Coordinates and provides appropriate assistance to the design organization with respect to preparation of necessary design review data (in other words data needed for the review)

9. Whenever proceedings of a design review allow, revises the system definition documentation

15.4.3 Useful Lead-in Items to the Design Review

According to Hill in Ref. 1, the items shown in Fig. 15.1 are effective lead-ins to the review.

15.5 DESIGN REVIEW TEAM

Reviewing the design of an item is not a one-person task; various people participate in the design review. The primary objective of all these personnel is to produce the best design at the minimum cost. This section discusses design review team members and their responsibilities. The following people usually form a typical engineering design review team:

1. *Design engineer or engineers*: These professionals prepare the design, make its presentation, etc.
2. *Senior design engineer or engineers*: These persons are not involved with the design under review but possess experience in the type of design under consideration. Basically they act as consultants.
3. *Procurement engineer or representative*: The basic responsibility of this person is to assure the availability of acceptable parts and materials on time at the best cost.
4. *Reliability engineer*: The basic function of this professional is to

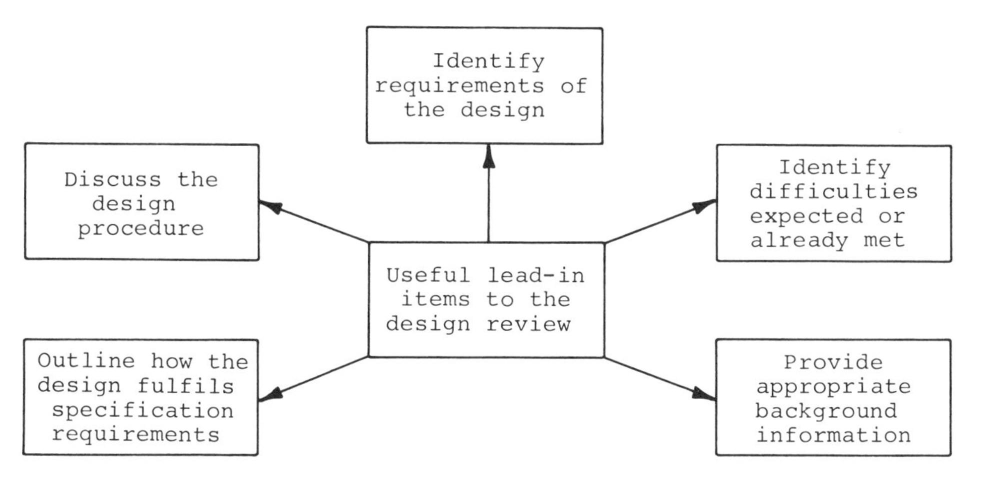

Figure 15.1 Useful lead-in items to the design review.

examine the design for optimum reliability and assure that the design meets the specified reliability requirements.

5. *Tooling engineer*: The primary responsibility of the tooling engineer is to examine the design in terms of tooling costs.
6. *Quality control engineer*: This person evaluates the design with respect to quality—for example, ensuring that the control, inspection and testing tasks can be performed effectively.
7. *Materials engineer*: The sole responsibility of this specialist is to ensure that the chosen materials will perform according to specifications or requirements.
8. *Manufacturing engineer*: This person evaluates the design so that it is producible at the lowest cost and in a specified time.
9. *Packaging and shipping representative*: This specialist examines the design from the packaging and shipping aspect, in other words, making sure that the resulting product can be handled effectively.
10. *Field engineer*: The field engineer evaluates the design from the aspects of maintenance, installation, and so on.
11. *User representatives*: Usually various specialists of the customer sit in on design reviews. The main purpose of these people is to ensure that the requirements of the customer are fully satisfied.
12. *Test engineer*: The test engineer examines the design from the various aspects of testing.
13. *Design review board chairman.*

According to Hill in Ref. 1, the number of persons participating in a design review should be no more than 12. More people will reduce the effectiveness of such a review. More information on the design review team is given in Refs. 2, 3, and 5.

15.6 RELIABILITY CONSIDERATIONS DURING THE DESIGN REVIEW

Attention is to be given to various reliability considerations during design reviews. Some of them are as follows:

1. Results obtained through testing
2. Reliability prediction results, calculations, and sources of data
3. Operational environments of the device
4. Failure of definitions
5. Reliability improvement plans
6. Plans for solutions to problems
7. Failure modes and effects analysis

8. Determination of tradeoffs between cost, performance, weight, maintainability, and so on, carried to optimize design
9. Requirements associated with performance

15.7 SOFTWARE DESIGN REVIEW

Today computers are commonly used in the newly designed engineering systems. This has led to the problem of quality and reliability of computer programs or software. Thus the review of the software design is an important component of the overall system design review.

According to Boehm in Ref. 7, it is in the initial stages of the software development process that the majority of the errors of the large software systems are introduced. In addition, prior to coding about 60 percent of total software errors are introduced. During the review of software design, many of such errors can be identified and corrected before the start of coding.

15.7.1 Types of Software Design Reviews

Software design reviews are classified into three categories [8]. These are as follows:

1. System design review
2. Software preliminary design review
3. Software critical design review

In the system design review the technical requirements allocated to software are evaluated. This evaluation is carried out after the documentation of such requirements in the computer program performance specification.

The software preliminary design review is conducted after defining the software to the computer program component level. Before the detail design the software preliminary design review evaluates the chosen software design procedure. In addition, during the software preliminary design review the functional interfaces are examined. Finally, the software critical design review is concerned with evaluating, before the start of coding, the fully completed detail design.

15.8 SUMMARY

This chapter is concerned with various aspects of engineering design reviews. Three types of design reviews are described. These are preliminary design review, intermediate design review, and critical design review.

The subjects discussed during design reviews are briefly explained. These are electrical, mechanical, specifications, standardization, reliability, drafting, human engineering, reproducibility, maintainability, value engineering, safety, and finishing.

Items of data required for design reviews are listed. Qualities and functions of the design review board chairman are presented.

Design review team members and their responsibilities are discussed. These members are the design engineers, procurement engineer, reliability engineer, tooling engineer, quality control engineer, materials engineer, manufacturing engineer, packaging and shipping representative, field engineer, customer representatives, test engineer, and design review board chairman.

Reliability considerations during the design reviews are listed. Finally the chapter describes the software design reviews. Three types of software design reviews are briefly described. These are system design review, software preliminary design review, and software critical design review.

EXERCISES

1. Explain in detail the purpose of design reviews.
2. Explain the following terms:
   a. Specification review
   b. Mechanical design review
   c. Final design review
   d. Preliminary design review
   e. Intermediate design review
   f. Electrical design review
3. List at least ten questions which can be asked of the designer of the product during the design review.
4. What information should be made available to the design review team?
5. What are the tasks of the design review board chairman?
6. List the typical characteristics of the design review board chairman.
7. What is the principal difference between the hardware design review and the software design review?
8. Who are the members of the design review team?
9. What are the functions of the design review team members?

REFERENCES

1. P. H. Hill, *The Science of Engineering Design*, Holt, Rhinehart and Winston, New York, 1970, pp. 239–242.
2. C. L. Carter, The Control and Assurance of Quality Reliability and Safety, C. L. Carter, Jr. & Associates, Inc., Management and Personnel Consultants, P.O. Box 50001, Richardson, Texas, 75080, 1978, pp. 301–312.

3. D. P. Simonton, Way a Design-Review Committee Pays Off Dividends, in *Management Guide for Engineers and Technical Administrators*, edited by N. P. Chironis, McGraw-Hill, New York, 1969, pp. 234–236.
4. Engineering Design Handbook, Development Guide for Reliability, Part Two, Design for Reliability. AMCP 706-196. Published by Headquarters, U.S. Army Material Command, 5001 Eisenhower Avenue, Alexandria, Virginia 22333, 1976, pp. 11.1–11.10.
5. R. S. Cazanjian, Design Practices and Review Procedures for Reliability, in *Reliability Engineering for Electronics Systems*, edited by R. H. Mayers, K. L. Wong, and H. M. Gordy, John Wiley & Sons, New York, 1964, pp. 152–171.
6. Handbook of Reliability Engineering, NAVWEPS 00-65-502, published by the Director of the Chief of the Bureau of Naval Weapons, 1964, pp. 3.2–3.3.
7. B. W. Boehm, Software Engineering, *IEEE Trans. Comput.,* Vol. 25 (December 1976), pp. 1226–1241.
8. J. McKissick, Quality Control of Computer Software, Proceedings of the American Society for Quality Control Technical Conference, 1977, pp. 391–398.

16

Reliability and Maintainability in Systems Design

16.1 INTRODUCTION

Considerations given to reliability and maintainability during the system design phase play an important role in the resulting reliability of the system in the field environment. Obviously if the system is designed without performing any reliability or maintainability analysis it is quite unlikely that the final system will be highly reliable. Therefore, it is not incorrect to say that the effort put into the reliability and maintainability aspects during the design stage will be the determining factor of product reliability in the field environment.

During the design phase of the project, the cost of reliability and maintainability analysis will increase the price of the system or product, but the field support cost of the product in question will theoretically be lower. Thus the cost factor plays a significant role in the amount of effort diverted to reliability and maintainability at the design stage. In some situations, the customers are willing to pay a higher acquisition price for the equipment so that the full consideration is given to reliability and maintainability during the design phase. Such actions result from the belief that the field support cost of the equipment over its entire life will be lower with proper reliability and maintainability considerations at the design stage.

Therefore, this chapter addresses the various aspects of system design reliability and maintainability.

16.2 ACTIVITIES ASSOCIATED WITH RELIABILITY DESIGN

This section is concerned with the reliability design activities. These activities may vary from one project to another. Some of them are listed below [1]:

1. Load and stress analysis
2. Reliability prediction
3. Failure modes and effects analysis
4. Reliability allocation
5. Trade-off studies
6. Burn-in evaluation
7. Preparation of critical parts list
8. Testing and reliability growth monitoring
9. Reviewing design and changes

16.3 RELIABILITY IMPROVEMENT CONSTRAINTS AND PRODUCT RELIABILITY MODIFICATION

This section addresses two areas. These are

1. Reliability improvements constraints
2. Product reliability modification

Both these items are discussed separately below.

16.3.1 Reliability Improvement Constraints

There are various constraints to which reliability improvement is subjected. For example, efforts to improve reliability are constrained by factors such as the following [2]:

1. Weight of the end product
2. Logistics and maintenance philosophy
3. Time schedules
4. Limitations associated with operator training
5. Cost of effort associated with design
6. Availability of personnel to perform the task
7. Cost associated with components manufacture
8. Unclear operational environments
9. Resistance of customer to certain configurations
10. Poor knowledge of properties of parts or materials
11. Procured parts availability

16.3.2 Product Reliability Modification

This section is concerned with those areas by changing which the product reliability can be modified. Thus the reliability of a product can be altered by changing, for example,

1. Specific parts
2. Materials used
3. Mission success definition
4. System configuration
5. Specific subsystems
6. Details of manufacture
7. Operating procedure

16.4 FUNDAMENTAL PRINCIPLES OF RELIABILITY DESIGN

The designer should be aware of basic fundamental principles of reliability design such as the following [3]:

1. Make use of proven parts as much as possible.
2. Constantly review and update existing standards.
3. Simplify item configuration.
4. Make use of redundancy whenever it is necessary.
5. Identify and eliminate critical failure modes with the aid of failure modes and effects analysis approach.
6. Evaluate tolerance with the aid of methods such as statistical tolerance analysis and worst case tolerance analysis.
7. Make use of mean life ratio approach when evaluating the life of a new item with respect to the old.
8. Make use of self-healing devices whenever it is possible. An example of a self-healing device is an automobile tire. In this case small punctures are automatically sealed by the layer of sealing compound.
9. Make use of preferred design concepts
10. Give each component enough strength so that is encounters worst stress without any difficulty. To enhance reliability the designer uses procedures such as decreasing variation of stress, increasing average strength, decreasing variation of strength, and decreasing average stress.
11. Introduce methods, devices, or both to detect impending failures.
12. Design for the longest possible period between preventive maintenance by keeping in mind the availability of personnel, overall maintenance policy, and accessibility.

13. Allocate and predict reliability.
14. Give consideration to factors such as human reliability, man–machine interface, and human factors.

16.5 DESIGN RELIABILITY ASSESSMENT PROCEDURE

This approach is concerned with assessing the reliability of the product during the design phase. The procedure is composed of eight steps [4,5]. All of these steps are shown in Fig. 16.1. Each of the steps in Fig. 16.1 is described below.

Establish System Definition

This is concerned with defining the system—for example, defining physical constraints, duty cycles, boundary conditions, environmental factors, operating modes, criteria of success and failure, mission profiles, and so on. The functional block diagram of the system helps in establishing the system definition.

Draw Reliability Block Diagram of the System

In this step the reliability block diagram of the system is developed. Both the reliability and functional block diagrams are very similar. The reliability block diagram is the modified version of the functional block diagram where those areas are stressed which influence reliability.

Determine the Quantity of Components in Each Block of the Reliability Block Diagram

This step is concerned with preparing a list of parts in each block.

Obtain Necessary Data for Each Component

This step is concerned with collecting necessary data for each component. The data sources are the military handbooks, users' data banks, handbooks prepared by the professional societies, etc.

Calculate Failure Rate of Each Component

In this step, the failure rate of each component is computed with the aid of collected data.

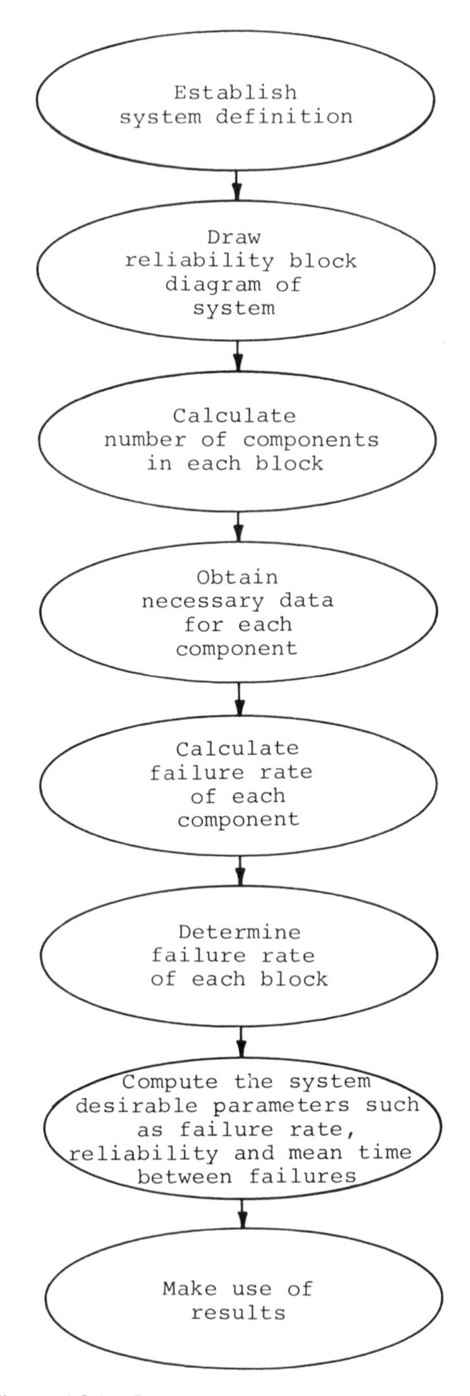

Figure 16.1 Design reliability assessment procedure.

Determine Failure Rate of Each Block

From the earlier step the failure rate of each component of each block is known. Thus, the next logical step is to combine the failure rates of components to compute the failure rate of each block. In this step the failure rate of each block is computed.

Compute the System Failure Rate, Reliability, and Mean Time Between Failures

With the aid of results of the previous step, in this step the values of the desirable parameters of the system are computed, for example, the system reliability, failure rate and mean time between failures.

Make Use of Results

The end results help to identify the weak areas of the system, the extent of improvement required, etc. Thus this step is concerned with taking appropriate actions as indicated by the end results.

16.6 DESIGN ENGINEERS

Design engineers play an important role with respect to reliability in system design. Poorly trained and poorly motivated design engineers from the reliability viewpoint will not be effective in designing a reliable system. Therefore this section discusses the problems of motivating design engineers for reliability and improving the effectiveness of designers [6–8]. Both these topics are described below, separately.

16.6.1 Difficulties in Motivating Design Engineers for Reliability

These difficulties are outlined in Fig. 16.2. All these activities are considered to be self-explanatory; therefore, their description is omitted. However, all these problems are described in Ref. 6.

16.6.2 Enhancing Effectiveness of Engineering Designers

The effectiveness of designers can be improved by bringing them to a state of self-sufficiency. This is accomplished through training and design experience retention [4].

The training helps designers to become proficient in the use of tools such as design of experiments and reliability quantification.

Similarly, with respect to design experience retention, it is an accepted fact that experience is an important need of a competent designer. The tools

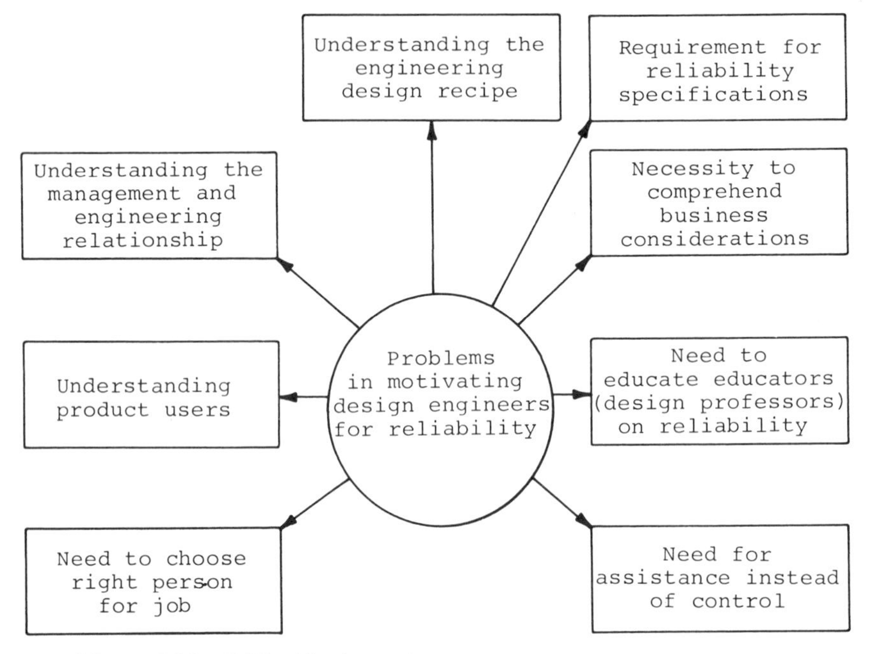

Figure 16.2 Difficulties in motivating design engineers for reliability.

used to achieve experience retention are checklists, failure data banks and their usage, manual of design standards, and so on.

16.7 MAINTAINABILITY CONSIDERATIONS IN PRODUCT DESIGN

During the product design phase, attention is also given to maintainability. If no attention is paid to maintainability during the design phase either the user of the product has to pay high cost of maintenance during the useful life period of the product or the manufacturer has to make expensive design changes in later stages of the product development cycle. However, during the design phase either the design engineer or the maintainability engineer obtains solutions to questions such as the following [8]:

1. What is the reason for designing the product and what are its operational requirements?
2. What are the maintenance objectives?
3. What are the policy considerations?
4. What are the environmental considerations?

5. Who is to support the product and how can the product be supported?
6. Where is the product to be supported and when can it be maintained?

The above questions can be answered with the aid of field failure data on similar products or systems. According to Ref. 8 items such as those given below should be included in the maintainability design criteria:

1. Reduction of maintenance downtime hours and complexity of maintenance
2. Reduction of maintenance error occurrence potential
3. Reduction of maintenance support costs and maintenance activities frequency
4. Limitation of maintenance personnel needs

16.7.1 Maintainability Design Characteristics

This section lists basic system characteristics which should be given careful attention during design for maintainability. Some of them are as follows [8]:

1. Accessibility, operability, and modular design
2. Standardization, interchangeability, and ease of removal
3. Size, shape, and weight
4. Manpower skills, tools, test equipment, and training needs
5. Displays, controls, calibrations, and labeling
6. Lubrication and installation
7. Wiring and interconnecting wires
8. Test points, hookups, and adapters
9. Illumination

16.7.2 Design for Accessibility

In Ref. [8] the accessibility is stated as the relative ease with which an item can be reached for maintenance. Inaccessibility to an item frequently results in poor maintenance. Some of the factors which affect accessibility are as follows:

1. The number of times the access is to be entered
2. Visual needs of the maintenance worker when working on the task
3. The type of tasks to be performed through the access
4. The time specified to perform the maintenance tasks through the access

5. The clothing requirement of the maintenance workers
6. Safety requirements associated with the use of the access
7. The distance that must be reached within the access by the maintenance worker
8. Operational location of the item in question

16.7.3 Modularization in Design

The modularization is concerned with the division of an item into functionally and physically distinct parts to facilitate replacement and removal. The following are some of the benefits of the modular construction:

1. Relatively low skill levels are required to replace modules in the field.
2. Relatively few tools are needed to replace modules in the field.
3. Divisible configuration is easier to maintain.
4. Repair can be accomplished more efficiently.
5. User maintenance personnel can be trained in lesser time.
6. Fully automated methods can be used to manufacture the standard "building blocks."
7. Use of standard "building blocks" helps to simplify new system design. Furthermore the design time is reduced.

16.7.4 Standardization in Design

Standardization is concerned with restricting to a minimum the variety of components that will satisfy the maximum of the system needs. Whenever the system design configuration is considered, the standardization must be a basic objective. The standardization has many objectives, some of which are as follows:

1. Interchangeable components should be used as much as possible.
2. Minimize the use of a number of different models in the field.
3. Increase the usage of standard off-the-shelf parts.
4. Increase the usage of common components in different products.
5. Simplify the problems of storage.

The followings are some of the advantages of standardization:

1. Reduces the procurement cost of a system
2. Reduces the training needs for support personnel
3. Reduces the support cost of a system over its entire life cycle
4. Increases the system reliability
5. Reduces the needs for technical publication
6. Increases the system maintainability

7. Reduces the requirement for support facilities
8. Reduces the varieties of test equipment needed to support a system
9. Reduces the amount of test equipment needed to support a system

16.8 SUMMARY

This chapter explores various aspects of reliability and maintainability in systems design. The reliability design associated activities are listed. Factors which constrain efforts to improve reliability are presented.

The 14 fundamental principles of reliability design are discussed. A procedure to assess design reliability is described. The procedure is composed of eight steps.

The next topic discussed in the chapter is design engineers. Two areas concerning design engineers are discussed: These are associated with problems in motivating design engineers for reliability and improving effectiveness of engineering designers.

Some aspects of maintainability considerations in product design are described. These are maintainability design characteristics, accessibility modularization, and standardization. Several maintainability design characteristics and factors which affect accessibility are specified. Advantages of the modular construction are presented. Five goals of standardization and its advantages are outlined.

EXERCISES

1. What are benefits of modularization and standardization in engineering design?
2. What are the goals of standardization?
3. How would you motivate engineers during the design phase for reliability and maintainability?
4. How would you change the reliability of a system at the design stage? Give at least eight distinct ways.
5. What are the design-related reliability and maintainability activities?
6. Describe an approach to predict reliability of a system during its design phase.
7. Discuss the functions of the following specialists during the system design phase:
   a. Design engineer
   b. Reliability engineer
   c. Maintainability engineer

REFERENCES

1. D. G. Dickey, Developing Your Commercial Reliability Program, Proceedings of the American Society for Quality Control Conference, 1980, pp. 724–728.
2. Engineering Design Handbook: Development Guide for Reliability, Part II Design for Reliability, AMCP 706-196; published by Headquarters, US Army Material Command, 5001 Eisenhower Avenue, Alexandria, Virginia 22333, 1976, pp. 1.8–1.9.
3. Quality Assurance, AMCP 702-3, published by Headquarters, US Army Material Command, 5001 Eisenhower Avenue, Alexandria, Virginia 22333, 1968, pp. 5.1–5.9.
4. J. M. Juran, F. M. Gryna, and R. S. Bingham, editors, *Quality Control Handbook*, McGraw-Hill, New York, 1979, pp. 8.17–8.21, 8.64–8.65.
5. Handbook: Reliability Engineering, NAVWEPS 00-65-502, published by the Direction of the Chief of the Bureau of Naval Weapons, available from the Superintendant of Documents, U.S. Government Printing Office, Washington, D.C. 20402, 1964, pp. 5.1–5.18.
6. H. J. Bajaria, Motivating Design Engineers for Reliability, Proceedings of the American Society for Quality Control Conference, 1979, pp. 767–773.
7. H. J. Bajaria, Motivating Design Engineers for Reliability, Part II, Proceedings of the American Society for Quality Control Conference, 1980, pp. 168–176.
8. Engineering Design Handbook: Maintainability Engineering Theory and Practice, AMCP 706-133, published by Headquarters, US Army Material Command, 5001 Eisenhower Avenue, Alexandria, Virginia 22333, 1976, p. 5.21.

17

Human Factors in Engineering Design

17.1 INTRODUCTION

Any product the engineer builds is ultimately for use by people. Therefore, people's needs, desires, and limitations have to be clearly understood by the engineer. Furthermore, all these things have to be considered during product design. This way only those engineering products will be produced which satisfy people's requirements.

The objectives of the human factors program are to improve comfort, human performance and use of personnel, reduce loss time due to human-related accidents, reduce training cost, and so on [1]. Human engineering is defined as the adaptation of human tasks to work environments and the working environment to human attributes such as mental, perceptual, sensory, and physical [2]. This adaptation for human use is applicable to various functions, for example, work method development, equipment design, and consumer products. Many disciplines have been involved in human engineering. Some of them are psychology, statistics, mechanical engineering, electrical engineering, and industrial engineering [3].

This chapter describes various aspects of human factors and related areas in engineering design.

17.2 HUMAN ELEMENT CONSIDERATIONS DURING DESIGN PHASE

This section briefly describes the considerations given to the human element during the three stages of system development. These are

1. Conceptual design stage
2. Preliminary design stage
3. Detailed design stage

Each of the above stages is described below with respect to the human element.

Conceptual Design Stage

During this stage it is expected that the design engineer or human factors engineer should give attention to items such as the following [4]:

1. *Potential user characteristics*: for example, size, intelligence level, age, reaction time, physiological tolerance, sex, health, mobility, and training.
2. *System features*: for example, task and operational stresses, hazards, procedures and instructions for the task, reach and clearance, and environmental stresses.
3. *Defining preliminary task descriptions*: for example, of user, operator, and maintainer.

Preliminary Design Stage

During this phase of the design the design engineer reviews the previous results. In addition, the design engineer or the human factor engineer conducts studies or analyses such as the following:

1. Man–machine simulation and mockup studies.
2. Time line analysis, which is basically concerned with finding out the feasibility of planned use of the operator or maintainer.
3. Refinement of task analysis.
4. Link analysis, which is concerned with examining the management of human tasks in relation to logical grouping and sequencing of activities.
5. Refinement of task description.

Detailed Design Stage

During this phase, some of the human engineering activities to be performed are as follows:

1. Safety and hazard analysis;
2. Time-line and link analyses;
3. Personnel requirement analysis: this is concerned with determining the number and type of persons needed to staff the system in the field;

4. Task and training requirements analyses;
5. Technical literature publications: these publications should take into account the human element—in other words, these publications should be prepared in such way that they can be used effectively by the user.

17.3 GUIDELINES FOR MONITORING THE HUMAN ENGINEERING ASPECTS OF DESIGN

To monitor the human engineering aspects of design, the various guidelines are specified in Fig. 17.1.

This section briefly lists those areas which are to be given appropriate attention when considering human factors during design trade-off studies. Some of these areas are as follows [4]:

1. Cost of personnel
2. Human performance reliability
3. Implications associated with training
4. Safety
5. Probability of acceptance of alternative designs by operators or maintainers
6. Human energy cost

17.4 SENSORY CAPACITIES OF HUMANS

This section briefly describes human sensory capacities. The description is concerned with sight, touch, noise, vibration, and motion. Understanding of such items is essential for the designer to produce an effective design [5].

Touch Capacity

Touch is an essential human sensor. People's capability to interpret auditory and visual stimuli is related to the sense of touch. Touching relieves the load on eyes and ears to convey messages to the brain. For example, control knob shapes can be distinguished by touching. Therefore, a designer can take advantages of the touch capacity of humans in situations where users have to depend completely upon their senses of touch.

Noise Capacity

Noise is a contributory factor to human feelings such as irritability, well-being, and boredom. Furthermore, noise may affect the quality of output generated by tasks requiring intense concentration. Sound levels above 130

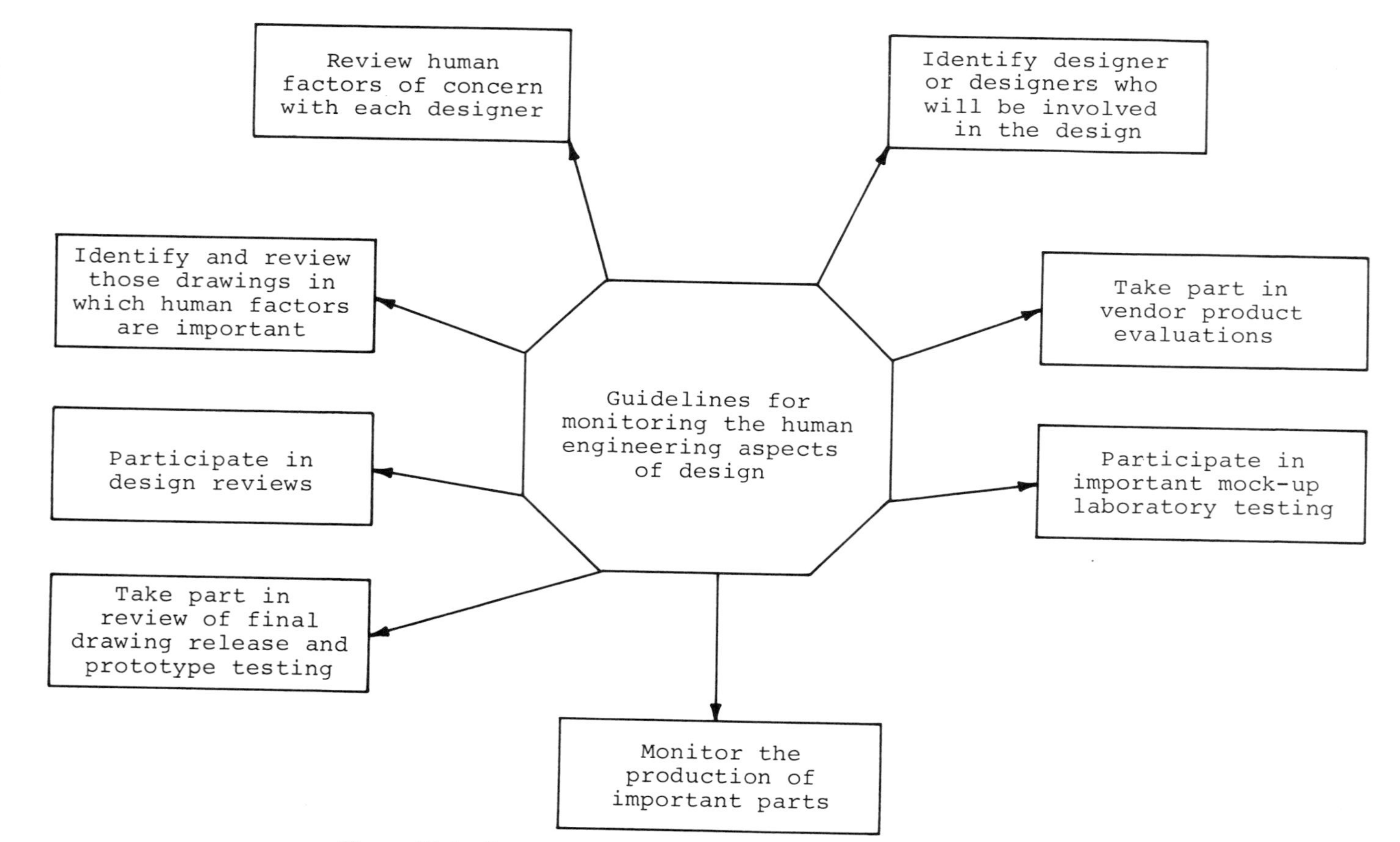

Figure 17.1 Suggestions to monitor the human engineering aspects of design.

dB become painful and unpleasant to humans. Excessive noise can damage hearing and make oral communication among operators and maintainers impossible. A noise level more than 100 dB is considered to be unsafe [5]. Furthermore, generally a level of noise below 90 dB is considered to be harmless.

Vibration and Motion Capacity

Vibrations may be a contributory factor to the poor performance of mental and physical tasks performed by operators and maintainers. For example, headaches, motion sickness and eye strain are contributed to by low-frequency, large-amplitude vibrations. To reduce the effects of vibration and motion in the equipment some useful guidelines are as follows:

1. Make use of springs, shock absorbers, etc.
2. In the seating design avoid 3 to 4 cycles/sec frequency vibrations (i.e., the resonance frequency of a person's vertical trunk when seated is between 3 and 4 cycles/sec).
3. Remember that the vertical vibrations affect seated people most.
4. To reduce vibrations make use of damping materials or cushioned seats.
5. Ensure that vibrations with amplitude in excess of 0.08 mil are not present.

Sight Capacity

A person's eye sees differently from different positions or angles. When a person looks straight ahead, he or she can perceive all colors; however, his or her perception starts to decrease with the increase in the viewing angle. For two different situations the limits of color vision are as follows:

1. *Vertically*: Green, 40°; yellow, 95°; white, 130°; blue, 80°; and red, 45°
2. *Horizontally*: Blue, 100°; white, 180°; yellow, 120°; and green-red, 60°

In the dark or at night small warning lights of yellow, blue, orange, and green color cannot be distinguished from a distance. All of them appear to be white in color.

17.5 AUDITORY AND VISUAL PRESENTATIONS

Sometimes in equipment design the designer has to make the decision whether to use auditory or visual presentation of a message. Therefore, this section describes separately situations where auditory and visual presentations are to be used [6].

Auditory Presentation

This is to be used in situations such as the following:

1. The message is short or it calls for a quick action.
2. The information to be presented is simple and straightforward.
3. The place where the information to be received is too bright.
4. No reference will be made to the message at a later stage.
5. The visual system of the person who will receive the information is already overburdened.
6. The tasks assigned to a person require him or her to move about on a continuous basis.

Visual Presentation

This is to be used in situations such as the following:

1. The place where the information is to be received has a noise level well above normal.
2. The size of the message is long.
3. Reference to the message will be made at a later stage.
4. The information to be presented is of complex nature.
5. The information to be presented does not call for quick action.
6. The auditory system of the person who will receive the information is already overburdened.
7. The tasks assigned to a person require him or her to remain in one position.

17.6 COMPARISON OF HUMAN AND MACHINE

Sometimes in engineering design the decision has to be made whether a task should be performed by a human or by a machine. Therefore this section briefly compares the capabilities and limitations of a human and a machine. Both capabilities and limitations are discussed below, separately.

Capabilities

Some of the comparisons of human capabilities with machine alternatives are as follows [4].

Human	*Machine*
1. Human possesses a long-term memory for events which were related.	1. It becomes very costly for the machine to have the same capability.

Human	Machine
2. Whenever the need arises humans are able to overide their own actions.	2. A machine can do only what it was designed to do.
3. Humans possess a considerable tolerance for uncertainty.	3. A machine is considerably limited by uncertainty in input.
4. In terms of job performance the human can become considerably flexible.	4. Relatively, the machine is inflexible.
5. Under the transient overload condition the human can work.	5. The machine does not function under the overload condition.
6. Whenever an input signal is subjected to a considerable noise, the human has the ability to interpret it.	6. Generally the machine can function effectively only in a noise-free environment.

Limitations

As for capability some of the comparisons of human limitations with machine alternatives are as follows [4].

Human	Machine
1. The channel capacity of humans is limited.	1. There is no limit to machine channel capacity—it may be as much as is affordable.
2. High gravitational forces have adverse effect on humans.	2. Gravitational forces do not affect the machine.
3. Humans are subject to things such as disorientation and motion sickness.	3. The machine is free of such things.
4. Fatigue and boredom degrade the performance of humans.	4. Lack of calibration or wearing out is the only factor that degrades the performance of the machine.
5. Humans are prone to factors such as anxiety.	5. The machine is free of such factors.
6. Due to interpersonal problems, humans are prone to stress.	6. The machine is free of this problem.

Human	Machine
7. Human beings do not always pursue the optimum approach.	7. The machine will always pursue the approach which it was designed to follow.
8. Humans have needs such as psychological and physiological ones.	8. In the case of machines, they have only the ecological need.

17.7 FACTORS SHAPING HUMAN PERFORMANCE

This section briefly describes those factors which shape human performance into an output. This output can either be identified as free of error performance or as an error. From Ref. 7, these factors are as follows:

1. *Situational characteristics*: These characteristics represent those job situations which are usually applicable to more than one task. Examples of the situational characteristics are air quality, work hours, noise and vibration, availability of supplies, humidity, and temperature.
2. *Psychological stresses*: These include uneventful long vigilance periods, threats associated with loss of job or failure, speed of the task, monotonous type work, task load, etc.
3. *Individual factors*: These are concerned with an individual worker. Examples of individual factors are attitudes, prior experience, physical condition, motivation, and personality variables.
4. *Job instructions*: The quality of job instructions is very important to a worker unless he or she has developed the skill to a level where no such instructions may be needed. Job instructions include work methods, oral or written communications, warnings, procedures and approaches, etc.
5. *Physiological stresses*: These are concerned with the physiological aspect of a worker. Thus, they include insufficient oxygen, fatigue, pain, extremes of atmospheric pressure, and so on.
6. *Task and equipment characteristics*: Industrial tasks usually require interaction of the worker with some kind of machine. The interaction demands must not be allowed to exceed the worker's limitations and capabilities. In any case, task and equipment characteristics includes memory, perceptual requirements, decision making, factors associated with person–machine interface, etc.

17.8 USEFUL FORMULAS FOR ENGINEERING DESIGNERS

This section presents selective human factor related formulas.

Formula I

This formula is due to Murrell [8] and is concerned with computing the length of the rest (schedule or not scheduled) period for any specified job activity. Thus, the rest required in minutes is given by

$$t_r = \frac{t_{wt}(k_a - k_s)}{k_a - C} \tag{17.1}$$

where

t_r = the required rest time in minutes
k_a = the mean kilocalories per minute of work
t_{wt} = the total working time in minutes
k_s = the kilocalories per minute (these are adopted as a standard)
C = a constant whose value is estimated to be 1.5; this is the approximate resting level in kcal/min

According to Ref. 8, the value of k_s is usually taken to be between 4 and 5 kcal/min.

Example 17.1 The following values for k_a, t_{wt}, k_s, and C are specified:

$$k_a = 7 \text{ kcal/min}, \quad t_{twt} = 120 \text{ min}, \quad k_s = 5 \text{ kcal/min}$$
$$C = 1.5 \text{ kcal/min}$$

Calculate the required rest time in minutes.

Substituting the specified data into Eq. (17.1) we get

$$t_r = \frac{120(7 - 5)}{7 - 1.5} = 43.64 \text{ min}$$

Formula II

This formula is due to Frederick [9], who developed it to calculate the energy costs, E_c, in kcal/hr. The formula can be used for any weight, any number of lifts per hour, and for any lift range. Thus

$$E_c = (ELWh)(1000)^{-1} \tag{17.2}$$

where

E = the energy cost per lift (g cal/ft lb)
W = the weight (lb)

h = the lifting height (ft)
L = the number of lifts per hour

Example 17.2 The following values are specified for E, L, W, and h:

$$E = 5 \text{ g cal/ft lb}$$
$$W = 20 \text{ lb}$$
$$h = 1.5 \text{ ft}$$
$$L = 200 \text{ lifts/hr}$$

Compute the value of E_c.

By substituting the specified data into Eq. (17.2), we get

$$E_c = \frac{(5)(200)(20)(1.5)}{1000} = 30 \text{ kcal/hr}$$

Formula III

This formula was developed by Peters and Adams [10] for determining the required character height, c_h, when labeling panel fronts. The formula equation is defined as follows:

$$c_h = f_{iv} + f_{im} + (0.0022)d_v \tag{17.3}$$

where f_{im} is the correction factor for criticality (for critical and noncritical markings the value of f_{im} is taken as 0 and 0.075, respectively); f_{iv} is the correction factor for viewing and illumination conditions; its values are specified in Ref. 1; and d_v is the viewing distance in inches.

Values of c_h are tabulated in Ref. 1.

Formula IV

This is concerned with calculating the visual angle in minutes. The visual angle is important because the message legibility depends on the visual angle subtended at the person's eye by the viewed object [1]. Thus, the visual angle is given by [11]

$$A_v = (60)(57.3)H/d \tag{17.4}$$

where A_v is the visual angle in minutes, d is distance from the target, and H is the height of the character or the diameter of the target.

The constant value specified in Eq. (17.4) is applicable only for those angles which are less than 10°.

Formula V

This formula is used to calculate brightness contrast. This is expressed as follows:

$$\alpha = (A_b - A_d)(A_b^{-1})100 \tag{17.5}$$

where α is the brightness contrast, A_b is the luminance of the brighter of two contrasting areas, and A_d is the luminance of the darker of two contrasting areas.

Example 17.3 Assume that a certain paper has a reflectance of 85% and the print on the paper has a reflectance of 10%. Calculate the value of the brightness contrast.

Substituting the above data into Eq. (17.5) results in

$$\begin{aligned}\alpha &= (85 - 10)(85)^{-1}(100) \\ &= 88.24\%\end{aligned}$$

Formula VI

This formula is concerned with calculating the human reliability of a time continuous task. Examples of time continuous tasks are scope monitoring and aircraft maneuvering. The human reliability is expressed by the following relationship [12]:

$$R_h(t) = \exp\left[-\int_0^t e(x)\,dx\right] \tag{17.6}$$

where $R_h(t)$ is the human reliability at time t and $e(t)$ is the time-dependent human error rate.

Example 17.4 After performing failure data analysis, it was established that the constant error rate of a person performing a time continuous task is 0.005 errors/hour. Compute the human reliability for a one-hour mission.

In this example the data are specified for $e(t)$ and t. Thus

$$e(t) = 0.005/\text{hr} \qquad \text{and} \qquad t = 1 \text{ hr}$$

Substituting the given data into Eq. (17.6) yields

$$R_h(1) = e^{-(0.005)(1)} = 0.995$$

The human reliability is 99.5%.

17.9 SUMMARY

This chapter briefly addresses the subject of human factors in engineering design. Human factors considerations during the three phases of the design are discussed. These phases are conceptual design, preliminary design, and detailed design. Eight guidelines for monitoring the human engineering aspects of design are presented, and human factors during design trade-off studies are discussed. The next topic described in the chapter is sensory capacities of humans. The sensory capacities described are touch, noise, vibration and motion, and sight.

Some of the uses of auditory and visual presentations are outlined. Human capabilities and limitations are presented. Six human performance-shaping factors are described. These are situational characteristics, psychological stresses, individual factors, job instructions, physiological stresses, and task and equipment characteristics. Finally, the chapter presents six selective formulas which will be useful to engineering designers. These formulas are concerned with calculating the rest time, the energy costs in kcal/hr, required character height, the visual angle, the brightness contrast, and the human reliability.

EXERCISES

1. Describe the following terms:
   a. Human noise capacity
   b. Human sight capacity
   c. Luminance
   d. Illuminance
   e. Human error rate
2. What are the limitations of humans in relation to machines?
3. List at least eight uses of the auditory presentation system in comparison to visual presentation system.
4. How would you assure the proper human factor consideration in the engineering design?
5. Define the term "human reliability."
6. Human reliability is expressed by Eq. (17.6). Develop an expression for the mean time to error for Rayleigh distribution.
7. The error rate of a time continuous task is estimated to be 0.005 errors/hr. Determine the value of the mean time to error.

REFERENCES

1. R. D. Huchingson, *New Horizons for Human Factors in Design*, McGraw-Hill, New York, 1981, pp. 7, 129, 47.

2. E. J. McCormick, *Human Engineering*, McGraw-Hill, Book Company, New York, 1957, pp. 1.
3. D. H. Edel, editor, *Introduction to Creative Design*, Prentice-Hall, Englewood Cliffs, New Jersey, 1967, pp. 73–74.
4. W. E. Woodson, *Human Factors Design Handbook*, McGraw-Hill, New York, 1981, pp. 10–12.
5. Engineering Design Handbook, Maintainability Guide for Design, AMCP 706-134, published by Headquarters, US Army Material Command, Washington, D.C. 20315, 1972, pp. 9.2–9.8.
6. J. P. Vidosic, *Elements of Design Engineering*, The Ronald Press, New York, 1969, pp. 125–127.
7. A. D. Swain, Design Techniques for Improving Human Performance in Production, published by Industrial and Commercial Techniques, Ltd., 30 Fleet Street, London, EC4Y 1AD, 1973, pp. 15–31.
8. K. F. H. Murrell, *Human Performance in Industry*, Reinhold, New York, 1965, p. 376.
9. S. W. Frederick, Human Energy in Manual Lifting, *Mod. Mater. Handling*, Vol. 14 (1959), pp. 74–76.
10. G. A. Peters and B. B. Adams, Three Criteria for Readable Panel Markings, *Product Eng.*, Vol. 30 (1959), pp. 55–57.
11. W. F. Grethers and C. A. Baker, Visual Presentation of Information in Human Engineering Guide to Equipment Design, edited by H. P. Van Cott and R. G. Kinkade. Available from the U.S. Government Printing Office, Washington D.C., 1972, Chap. 3.
12. B. S. Dhillon, *Reliability Engineering in Systems Design and Operation*, Van Nostrand Reinhold Company, New York, 1983, p. 71.

18

Design Optimization Methods

18.1 INTRODUCTION

Since the development of the electronic computer, the subject of optimization has been receiving considerable attention and many new models and procedures have surfaced. Broadly speaking, optimization may be described as the search for the solution that will generate the "maximum benefit." Furthermore the optimal design may be defined as the best of all feasible designs. Always, the aim of both the design methodology and the design engineer is to seek an optimum solution to the design problem. According to Middendorf [1], the likelihood of developing an optimum design basically depends on whether a mathematical model is established or not. However, sometimes designers with substantial design experience may develop a near optimum design without having a mathematical model. Nowadays, various procedures and techniques are available to optimize a mathematical model. This chapter presents some of these techniques.

18.2 GENERAL FUNDAMENTALS OF OPTIMIZATION

This section briefly discusses general fundamentals of optimization. Some of them are as follows [2]:

1. Optimization is associated with independent systems only. In other words, only those systems can be optimized which are independent of all others.

2. In optimization “trade-off” decisions have to be made because usually desirable optimizing objectives conflict with some other objectives.
3. The objective of the engineer is to arrive at the optimum solution while meeting the specified needs and using limited resources.
4. The selected solution is the best technically conceivable but is never really optimum.
5. Perform optimum analysis when enough knowledge about the technical field has been gained and the field is well developed.
6. Because of unavoidable constraints associated with the system, entire subsystems of the systems are always suboptimized.

According to Ref. 2, the methods of optimization may be classified into the following four categories:

1. Graphical
2. Subjective
3. Analytical
4. Utilization of general principles

18.3 THE MOST GENERAL ANALYTICAL FORM OF THE OPTIMIZING PROBLEM

The problem is expressed as follows [3,4]:

$$A = f(y_1, y_2, y_3, y_4, \ldots, y_k) = \text{maximum or minimum} \tag{18.1}$$

$$G_j(y_1, y_2, y_3, y_4, \ldots, y_k) = 0 \qquad \text{for } j = 1, 2, 3, \ldots, n \tag{18.2}$$

$$d_i \leq g_i(y_1, y_2, y_3, y_4, \ldots, y_k) \leq D_i \qquad \text{for } i = 1, 2, 3, \ldots, m \tag{18.3}$$

where d_i and D_i are constants and $y_1, y_2, y_3, \ldots, y_k$ are the independent variables.

Equations (18.1), (18.2), and (18.3) are known as the value expression, the functional constraints, and the regional constraints, respectively [2].

18.4 OPTIMIZATION WITH DIFFERENTIAL CALCULUS

This is a well-known technique to engineers. The technique uses differential calculus to obtain optimum value. This technique calls for taking partial derivatives of a function with respect to each of its independent variables and equating to zero. For example,

$$\frac{\partial f(y_1, y_2, \ldots, y_k)}{\partial y_1} = 0 \tag{18.4}$$

$$\frac{\partial f(y_1, y_2, \ldots, y_k)}{\partial y_2} = 0 \tag{18.5}$$

$$\frac{\partial f(y_1, y_2, \ldots, y_k)}{\partial y_k} = 0 \tag{18.6}$$

where $f(y_1, y_2, \ldots, y_k)$ is a function of independent variables $y_1, y_2, \ldots, y_k$.

Equations (18.4)–(18.6) yield k simultaneous equations. These equations have to be solved for the optimum variables values. The sign of the second derivative of the function $f(y_1, y_2, y_3, \ldots, y_k)$ is the indicator whether the function is a maximum or minimum. Some of the drawbacks of this technique are the requirement for the existence of first and second derivatives of the function and finding a solution to a set of simultaneous equations [5]. For a single independent variable function, the following example is demonstrated.

Example 18.1 An electronic diode can fail either short-circuited or open-circuited. If d such independent and identical diodes are connected in parallel configuration, from Ref. 6, the reliability of such a system is given by

$$R = (1 - F_s)^d - F_0^d \tag{18.7}$$

where

F_s = the diode's short mode failure probability
F_0 = the diode's open mode failure probability
d = the number of diodes
R = the reliability of the parallel configuration

Determine how many diodes are to be connected in parallel for maximum reliability, if $F_s = 0.2$ and $F_0 = 0.3$.

Taking the partial derivative of Eq. (18.7) with respect to d results in

$$\frac{\partial R}{\partial d} = A^d \ln A - F_0^d \ln F_0 \tag{18.8}$$

where

$$A \equiv (1 - F_s) \tag{18.9}$$

Setting the right-hand term of Eq. (18.8) equal to zero yields

$$A^d \ln A - F_0^d \ln F_0 = 0 \quad (18.10)$$

Solving for d we get

$$d' = \ln[(\ln F_0)/\ln A]/\ln(A/F_0) \quad (18.11)$$

where d' is the optimum value.

By substituting the specified data into Eq. (18.10), we get

$$d' = \ln[\ln(0.3)/\ln(0.8)]/\ln[(0.8)/0.3)] = 1.72 \simeq 2$$

where

$$A = 1 - 0.2 = 0.8$$

Thus two diodes are to be connected in parallel in order to obtain maximum reliability from the configuration.

18.4.1 Constrained Optimization by Substitution and Differential Calculus

In the above example, the unconstrained optimization with differential calculus was described. This section is concerned with constrained optimization by differential calculus. In other words, in this case the restrictions have to be met while optimizing the specified function. The function to be optimized is known as the criterion function or objective function.

The method is demonstrated in the example below [7].

Example 18.2 Maximize

$$P_r = R_t - c_t \quad (18.12)$$

$$R_t \equiv \theta q_s \quad (18.13)$$

$$c_t \equiv c_f + c_v q_s \quad (18.14)$$

subject to

$$q_s = \alpha - \beta\theta \quad (18.15)$$

where

P_r = the profit
R_t = the total revenue
c_t = the total cost
θ = the price
q_s = the quantity sold

c_f = the fixed cost
c_v = the variable cost per unit
α and β = the demand function parameters

Equation (18.15) is the demand function. Obtain an expression for the optimum value of θ.

By substituting Eqs. (18.13)–(18.14) into Eq. (18.12) we get

$$P = \theta q_s - c_f - c_v q_s \tag{18.16}$$

Substituting the constraints (18.15) into Eq. (18.16) results in

$$\begin{aligned} P_r &= \theta(\alpha - \beta\theta) - c_f - c_v(\alpha - \beta\theta) \\ &= \theta\alpha - \beta\theta^2 - c_f - c_v\alpha + c_v\beta\theta \end{aligned} \tag{18.17}$$

Taking the derivative of Eq. (18.17) with respect to θ, we get

$$\frac{dP_r}{d\theta} = \alpha - 2\beta\theta + c_v\beta \tag{18.18}$$

Differentiating Eq. (18.18) with respect to θ results in

$$\frac{d^2P_r}{\partial\theta^2} = -2\beta \tag{18.19}$$

To find the maximum value of P_r, we set Eq. (18.18) equal to zero and solve for θ:

$$\theta' = (\alpha + c_v\beta)/2\beta \tag{18.20}$$

where θ' is the optimal price when the profit, P_r, is maximum.

18.4.2 Method of Lagrange Multipliers

This is another method which makes use of differential calculus. The method is used to find optima in multivariable problems containing constraints. Furthermore, it facilitates conversion of an objective or criterion function along with constraints into a single constraint-free function.

This function can be used to find the relative maximum or minimum values of the objective function. For example, to find relative maximum or minimum values of the objective function, $g(y_1, y_2, \ldots, y_k)$, subject to constraints $f_1(y_1, y_2, \ldots, y_k) = 0$, $f_2(y_1, y_2, \ldots, y_k) = 0$, $f_3(y_1, y_2, \ldots, y_k) = 0$, $f_m(y_1, y_2, \ldots, y_k) = 0$; we formulate a single unconstrained function $D(y_1, y_2, \ldots, y_k)$ as follows:

$$D(y_1, y_2, \ldots, y_k) = g + \lambda_1 f_1 + \lambda_2 f_2 + \lambda_3 f_3 + \cdots + \lambda_m f_m \tag{18.21}$$

Subject to the (necessary) conditions

$$\frac{\partial D}{\partial y_1} = 0 \tag{18.22}$$

$$\frac{\partial D}{\partial y_2} = 0 \tag{18.23}$$

$$\frac{\partial D}{\partial y_3} = 0 \tag{18.24}$$

$$\frac{\partial D}{\partial y_k} = 0 \tag{18.25}$$

where $y_1, y_2, y_3, \ldots, y_k$ are the variables and $\lambda_1, \lambda_2, \lambda_3, \ldots, \lambda_k$ are the Lagrange multipliers.

Example 18.3 A certain design problem is defined as follows:

$$\text{Maximize } z = 8x + 5y + xy - x^2 - y^2 + 12 \tag{18.26}$$

subject to

$$x + y - 7 = 0 \tag{18.27}$$

where x and y are the variables.

Maximize Eq. (18.26) by using the method of Lagrange multipliers.

From Eqs. (18.26)–(18.27) the single unconstrained function is formulated as follows:

$$\text{Maximize } D = 8x + 5y + xy - x^2 - y^2 + 12 - \lambda(x + y - 7) \tag{18.28}$$

where λ is the Lagrange multiplier.

Taking the partial derivatives of Eq. (18.28) with respect to λ, x, and y we get

$$\frac{\partial D}{\partial \lambda} = -x - y + 7 \tag{18.29}$$

$$\frac{\partial D}{\partial x} = 8 + y - 2x - \lambda \tag{18.30}$$

$$\frac{\partial D}{\partial y} = 5 + x - 2y - \lambda \tag{18.31}$$

By setting the right-hand term of Eqs. (18.29)–(18.31) equal to zero, results in the following three simultaneous equations:

$$-x - y + 7 = 0 \tag{18.32}$$

$$8 + y - 2x - \lambda = 0 \tag{18.33}$$

$$5 + x - 2y - \lambda = 0 \tag{18.34}$$

Solving the above three equations, we get

$$\lambda = 3$$

$$x = 4$$

$$y = 3$$

Thus, at $x = 4$ and $y = 3$, the value of z is at its maximum.

18.5 LINEAR PROGRAMMING

This is one of the widely used optimization techniques. It has been applied to various problems—for example, resource allocation, engineering design optimization, advertising, and system reliability optimization. Some examples of linear programming application to design problems are presented in Ref. 5.

The development of the method of linear programming is credited to George B. Dantzig. He published a paper containing the simplex method in 1947. According to Ref. 7, originally the linear programming method was known as "programming of interdependent activities in a linear structure." At a later stage the name was shortened to "linear programming." The term "linear" means directly proportional relationships between variables of the problem whereas the term "programming" applies to computing procedures associated with solving a system of linear equations [7]. Since 1947, various other researchers such as W. H. Cooper, A. Charnes, and A. Henderson have contributed to the linear programming method.

The linear programming problem may be expressed as follows:

Maximize (or minimize) $Y = c_1 z_1 + c_2 z_2 + \cdots + c_i z_i + \cdots + c_k z_k$ (18.35)

subject to

$$\sum_{i=1}^{k} b_{ji} z_i \leq a_j \qquad \text{for } j = 1, 2, 3, 4, \ldots, d \tag{18.36}$$

$$z_i \geq 0 \qquad \text{for } i = 1, 2, 3, 4, \ldots k \tag{18.37}$$

where

z_i = the ith variable for $i = 1, 2, 3, \ldots, k$
b_{ji} = the (j,i) constant, for $j = 1, 2, 3, \ldots, d$

a_j = the ith constant, for $j = 1, 2, 3, \ldots, d$
c_i = the ith constant, for $i = 1, 2, 3, \ldots k$

Equation (18.35) is known as the objective function whereas Eq. (18.36) is called constraint or constraints. Thus, it may be stated that linear programming is concerned with maximizing or minimizing the objective function subject to equality or inequality or both types of constraints.

The linear programming (LP) problem can be solved either graphically or analytically. The LP problem is solved graphically only if it has two variables; otherwise it is solved analytically. The simplex algorithm is used to solve a more complex linear programming problem. However, our discussion will be restricted to the graphical approach. The simplex method is described in detail in Ref. 7. The steps associated with the graphical approach are as follows:

Step 1: Identify the decision variables, the objective function, and constraints.
Step 2: Plot all constraints on the graph paper.
Step 3: Identify the feasible region on the graphical plot.
Step 4: Plot the objective function and choose the point on the feasible region that optimizes the value of the objective function.
Step 5: Interpret the resulting solution.

Example 18.4 Write down the formulation of the linear programming problem by setting $k = 3$ and $d = 4$ in Eqs. (18.35)–(18.37).

Thus from Eqs. (18.35)–(18.37) we get

$$\text{Maximize (or minimize) } y = c_1 z_1 + c_2 z_2 + c_3 z_3 \tag{18.38}$$

subject to

$$b_{11} z_1 + b_{12} z_2 + b_{13} z_3 \leq a_1 \tag{18.39}$$

$$b_{21} z_1 + b_{22} z_2 + b_{23} z_3 \leq a_2 \tag{18.40}$$

$$b_{31} z_1 + b_{32} z_2 + b_{33} z_3 \leq a_3 \tag{18.41}$$

$$b_{41} z_1 + b_{42} z_2 + b_{43} z_3 \leq a_4 \tag{18.42}$$

$$z_1 \geq 0 \tag{18.43}$$

$$z_2 \geq 0 \tag{18.44}$$

$$z_3 \geq 0 \tag{18.45}$$

Example 18.5

$$\text{Maximize } y = 2x + 3z \tag{18.46}$$

Subject to constraints

$$x + 2z \leq 12 \tag{18.47}$$

$$5x + 5z \leq 40 \tag{18.48}$$

$$x \leq 5 \tag{18.49}$$

$$x \geq 0 \tag{18.50}$$

$$z \geq 0 \tag{18.51}$$

The plots of constraints (18.47)–(18.51) are shown in Fig. 18.1. The shaded area IJKLM is the feasible solution region. This region is redrawn in Fig. 18.2. The plots of objective function (18.46) are shown for $y = 12$, 18, and 24 in Fig. 18.2. At $y = 12$ and 18 the plots of the objective function are well within the shaded feasible region. This indicates that the optimum value of y is greater than 18 but less than 24 as the objective function plot at $y = 24$ lies well outside the feasible region.

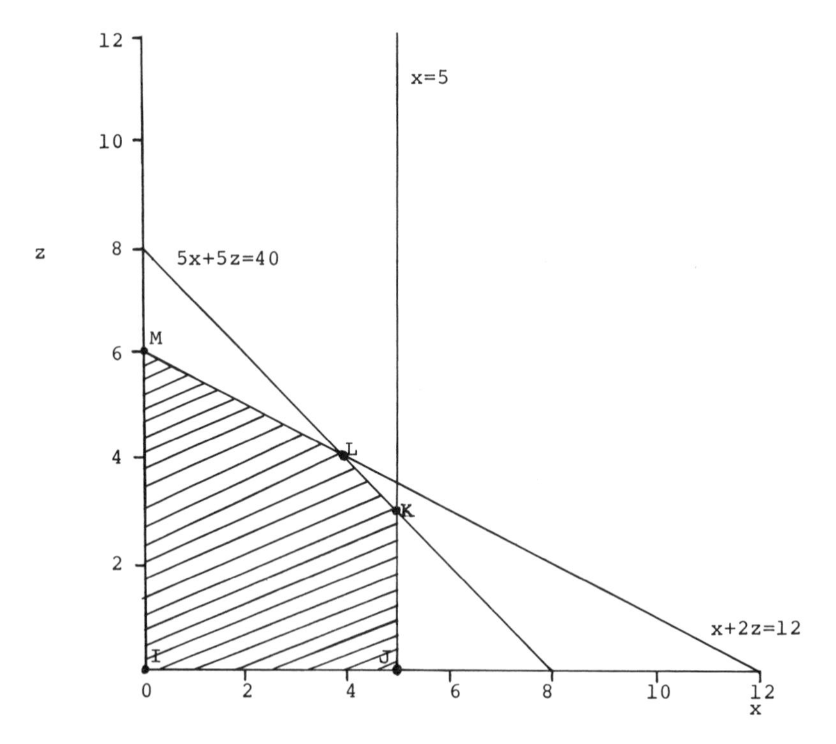

Figure 18.1 A graph of all constraints.

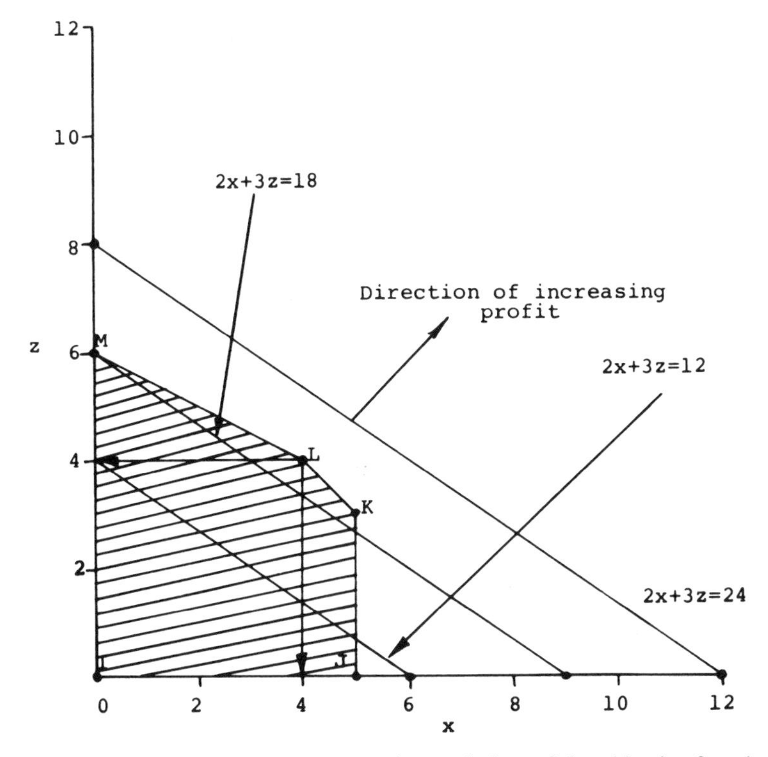

Figure 18.2 Feasible solution region and plots of the objective function.

These parallel lines indicate that L is the outermost point where a parallel line will still be within the feasible region. Therefore, it is the optimum point where the value of y will be at its maximum. At this point the values of z and x are equal to 4. Thus, by substituting both these values into Eq. (18.46) we get

$$y = 2(4) + 3(4) = 20$$

The maximum value of y is 20.

18.6 SUMMARY

This chapter presents selective techniques to optimize constrained and unconstrained design problems. The constrained design problem means that problem which is subject to some sort of constraint or constraints. Similarly, the unconstrained design problem means that problem which is free of constraints. General fundamentals of optimization are discussed along with the types of optimization procedures. The general analytic form of the optimizing problem is expressed. The optimization techniques which make use of differential calculus are presented. An example of a single variable design optimization problem is solved. This problem is concerned with the reliability optimization of a parallel configuration composed of identical electronic diodes. An example of a constrained optimization by substitution and differential calculus is presented. The method of Lagrange multipliers is described.

The last optimization technique covered in the chapter is known as linear programming. One example of the linear programming problem is solved graphically.

EXERCISES

1. Discuss the following three terms:
   a. Subjective optimization method
   b. Nonlinear optimization
   c. Constrained optimization
2. A series system is composed of k independent and identical electronic diodes. The reliability of the series system is given by

$$R = (1 - Q_0)^k - Q_s^k \tag{18.52}$$

   where R is the reliability of the series system, Q_s is the diode's short mode failure probability, and Q_0 is the diode's open mode failure probability. Develop an expression for the optimum number of diodes to be connected in series so that the series system reliability is at its maximum.
3. In Exercise 2, for $Q_0 = 0.1$ and $Q_s = 0.2$ calculate the maximum reliability of the series system.
4. Discuss the advantages and disadvantages of the method of Lagrange multipliers.
5. Compare the linear programming technique with the method of Lagrange multipliers.
6. Maximize

$$z = 3y_1 + 5y_2 \tag{18.53}$$

   subject to

$$y_1 + 2y_2 \leq 14 \tag{18.54}$$

$$4y_1 + 4y_2 \leq 40 \tag{18.55}$$

$$y_1 \leq 6 \tag{18.56}$$

$$y_1 \geq 0 \tag{18.57}$$

$$y_2 \geq 0 \tag{18.58}$$

REFERENCES

1. W. H. Middendorf, *Engineering Design*, Allyn and Bacon, Boston, 1969, pp. 184–185.
2. T. T. Woodson, *Introduction to Engineering Design*, McGraw-Hill, New York, 1966, pp. 258–283.
3. R. C. Johnson, *Optimum Design of Mechanical Elements*, John Wiley & Sons, New York, 1961.
4. J. N. Siddall, *Analytical Decision-Making in Engineering Design*, Prentice-Hall, Englewood Cliffs, New Jersey, 1972, p. 94.
5. G. Pitts, *Techniques in Engineering Design*, Butterworths, London, 1973, pp. 102–103.
6. B. S. Dhillon, The Analysis of the Reliability of Multistate Device Networks, available from the National Library of Canada, Ottawa, 1975.
7. S. M. Lee, L. J. Moore, and B. W. Taylor, *Management Science*, Wm. C. Brown Company, Dubuque, Iowa, 1981, pp. 756–758.
8. J. L. Riggs, *Production Systems: Planning, Analysis and Control*, John Wiley & Sons, New York, 1981, p. 165.

Index